VIRUSES

VIRUSES

BY

KENNETH M. SMITH

C.B.E., D.Sc., Ph.D., F.R.S.

Formerly Director, A.R.C. Virus Research Unit, Cambridge

CAMBRIDGE

AT THE UNIVERSITY PRESS

1962

PUBLISHED BY
THE SYNDICS OF THE CAMBRIDGE UNIVERSITY PRESS
Bentley House, 200 Euston Road, London, N.W. 1
American Branch: 32 East 57th Street, New York 22, N.Y.
West African Office: P.O. Box 33, Ibadan, Nigeria

Printed in Great Britain at the University Press, Cambridge
(*Brooke Crutchley, University Printer*)

CONTENTS

The half-tone illustrations (plates I–XVI) *are between pp.* 48 *and* 49.

I believe that the elucidation of nucleic acid in all its aspects is the most important scientific problem we face today. It is vastly more important than any of the problems associated with the structure of the atom, for in nucleic acid structure we are dealing with life itself and with a unique approach to bettering the lot of mankind on earth.

WENDELL M. STANLEY. *Penrose Memorial Lecture*, 1957

PREFACE

During a fairly long experience of working with and talking about viruses, one becomes familiar with certain questions which are always asked by the student and inquiring layman. Among these are such queries as: what do you mean by a 'virus', and how does it differ from other disease agents? What do viruses look like and how do they multiply? How can they be purified and crystallized, and, very frequently, do viruses cause cancer? These questions and others I have endeavoured to deal with in as interesting a manner as possible and with the minimum of technical description.

As to the question asked most frequently of all, 'Are viruses living organisms?', that must be left to the questioner himself to answer, after, as we hope, he has read this book.

Grateful acknowledgement is due to the editors of *Nature* and of *The Practitioner* for permission to reproduce the substance of the respective articles 'Recent Work on the Electron Microscopy of Viruses' and 'The Nature of Viruses'. The editor of *Outlook on Agriculture* kindly allowed me to reproduce the substance of an article from that journal, 'The Use of Viruses in the Biological Control of Insect Pests'. I am also grateful to those friends who lent me photographic prints for the illustrations; the names of these authors are given under each borrowed illustration. Finally, I must express my appreciation to Mr A. K. Parker of the Cambridge University Press for reading the text and for making many helpful suggestions.

KENNETH M. SMITH

AGRICULTURAL RESEARCH COUNCIL
VIRUS RESEARCH UNIT
CAMBRIDGE

1

INTRODUCTION

What is meant by the word 'virus'?

The word 'virus' comes from the Latin and means literally a poison; the *Oxford English Dictionary* describes it as a 'morbid poison, a poison of contagious disease, as smallpox'. The scientists who investigate viruses are known as 'virologists', a word of doubtful parentage begotten of a Greek and Latin *mésalliance*, but since no one could think of a better word this has now become established in everyday use.

After Louis Pasteur and Robert Koch had shown that infectious diseases were caused by minute living organisms or 'germs', it was confidently expected that the presence of such germs could be demonstrated for all infectious diseases. However, it soon became clear that this was not the case, and the newly developed technique of bacteriology, when applied to such diseases as mumps, measles or smallpox, failed to reveal any germ or bacterium which could account for the infections. Pasteur was equally unsuccessful with rabies and suggested the possibility of a submicroscopic organism, but the conception of a disease-agent of a completely different nature did not occur to him.

Discovery of the first virus

The first virus about which indisputable scientific evidence was obtained was one affecting the tobacco plant in which it caused a disease known as 'mosaic', a term now applied to many similar-appearing plant-virus diseases. Although the actual proof of the existence of such an agent was given by a Russian botanist, Ivanovski, in 1892 there were two other workers concerned with the investigation of this plant disease, and it is of interest to observe the reaction of these three when confronted with this new phenomenon.

Adolf Mayer, a German working in Holland about 1886, was the first to carry out any scientific studies of the tobacco mosaic disease. It was he who demonstrated that the sap from an infected tobacco plant reproduced the mosaic disease when injected into a healthy tobacco plant. Like Pasteur, who about this time was working with rabies, Mayer was unable to demonstrate the presence of any kind of disease agent. Nevertheless, Mayer was greatly influenced by the bacteriological advances of that time and he continued to look for a bacterium as the causal agent. In the meantime the tobacco-growers themselves had their own opinions as to what caused the mosaic disease and these were both numerous and bizarre. Some blamed the sun's rays, some considered the cold nights or fogs responsible, whilst others put the blame on the hotbeds in which the tobacco was grown. Many growers considered the disease to be entirely unexplainable and due to a kind of magic; they told Mayer that he would never find the cause. This conception of a virus as a mysterious almost supernatural entity persisted up to the late 1930's, when the pioneer work of Stanley in the U.S.A. and Bawden and Pirie in the United Kingdom was the first in the series of spectacular advances which constitute the epic of virus research over the last two and a half decades. Meanwhile Mayer, having decided that a bacterium must be the cause of the disease, cast his net far and wide. He inoculated his tobacco plants with a great number of well known bacteria and with all kinds of media containing bacteria. These latter varied from pigeon and other manures to putrefied legumes and old cheese. Needless to say none of these concoctions produced the mosaic disease in tobacco plants.

We come now to Ivanovski, who confirmed some of Mayer's findings but contradicted others. He agreed that the sap was infectious, that infectivity of the sap was lost by heating to near boiling-point, and that, in the absence of fungi and other parasites, infection by bacteria must be the cause of the disease. He contradicted most emphatically, however, Mayer's statement that the sap of leaves attacked by mosaic disease lost all infectivity after filtration through double filter-paper. Furthermore, he showed, and herein lies his claim to fame, that the sap was still infectious after passage of a Pasteur–Chamberland filter-candle which removed any visible organisms including all bacteria. However, like Mayer, Ivanovski still considered that bacteria were the cause of the disease but that they were of very small size and were submicroscopic. The third on the scene was a Dutch microbiologist, M. W. Beijerinck, who repeated and confirmed Ivanovski's experiments on the filter-passing capacity of the agent. Unlike the two previous workers, however, Beijerinck rejected the idea of a causal bacterium and conceived the idea of a *contagium vivum fluidum*. Although this conception is not very precise in its meaning it was the first step away from the conventional bacteriological approach to the problem.

Beijerinck was a careful investigator and, not being content with the results of the filtration experiments, carried out another test to convince himself that bacteria could not be concerned in the production of tobacco mosaic. He placed some sap from a mosaic-diseased plant on a thick agar plate and left it to diffuse for several days. His idea was that the 'agent', if capable of diffusion, would penetrate into the agar downwards and sideways, thereby leaving as a residue all discrete parts, aerobic and anaerobic bacteria and their spores. After the lapse of some days he removed the upper layer of agar and inoculated healthy tobacco plants with two successive underlying layers of agar. The inoculated plants duly developed the disease, and the success of this experiment convinced Beijerinck that 'the virus must really be regarded as liquid or soluble and not as corpuscular'. At about this time two German workers, Loeffler and Frosch, had shown that the foot-and-mouth disease of cattle was also caused by a similar filter-passing agent and Beijerinck finds that he 'cannot agree with Mr Loeffler as regards the corpuscular nature of the virus of foot-and-mouth disease'. We might perhaps state at this juncture, in order to set the mind of Beijerinck's ghost at rest, that the actual size of the particles in the *contagium vivum fluidum* of tobacco mosaic and foot-and-mouth disease are respectively 300 mμ. by 15 and 22 mμ (1 mμ, which is the symbol for millimicron, equals one millionth of a millimetre). We shall hear more about the sizes of viruses in chapter 4.

Viruses in history

Viruses are no modern phenomenon; they have existed for thousands of years and smallpox, a typical virus disease, was described by the Chinese in the tenth century before Christ. The spread of smallpox over the world was extremely slow, but now the disease has travelled the entire surface of the globe and is present wherever the human species exists. It has been estimated that in the eighteenth century 60 million people died of smallpox.

Yellow fever, or 'yellow jack' as it was called by sailors, has been known for centuries in tropical Africa as a scourge of ships in the African trade, and Bedson suggests that it was probably responsible for the legends of the cursed ships, the *Ancient Mariner* and the *Flying Dutchman*.

Yellow fever was confined to the Western Hemisphere before connexion between the two hemispheres was established by Columbus. During the French Revolution the quarantine of European ports was abolished, and the virus spread into Europe in the period 1791 to 1815, and to both North and South America.

The establishment of the Republic of Haiti was really brought about by yellow fever. After the defeat of the natives by the French army sent by Napoleon, the French were unable to follow up their victory owing to the ravages of yellow fever, and out of 25,000 men only 3000 survived to evacuate the island.

Owing to the confusion of influenza with other similar fevers it is not possible to trace it very far back in human history. It is, however, known to have been present in Europe during the sixteenth century. Great epidemics, or pandemics as they are called, seem to have occurred at intervals over a long period of years while, between the pandemics, lesser epidemics occur at intervals of two or three years. Between 1848 and 1889 there was a marked absence of influenza in England, but from 1890 onwards influenza has always been present.

One of the worst plagues in history was the great influenza pandemic of 1918–19, when about 150,000 people died in the United Kingdom alone. In most countries populated by Europeans the death-rate was 3–5 per 1000, while non-Europeans showed a much higher mortality; in India, for example, there were over 5 million deaths.

The earliest record of a plant-virus disease goes back to about the middle of the sixteenth century though, of course, no conception of such a thing as a virus existed at that time. This refers to a variegation in the flower-colour of tulips, now called a 'colour-break', and is caused by an aphid-transmitted virus. There are many of these 'broken' tulips illustrated in the paintings of Rembrandt and in old herbals of that date.

Some of these colours are extremely attractive, and in the early days of the tulip in Holland this led to a bizarre situation in which there was much financial speculation, and a bride was lucky if she could count a 'broken' tulip in her dowry.

Differences from other disease agents

It will be helpful if we give at the start a few of the fundamental characteristics of viruses. Though in some ways the agents classified as viruses are rather a heterogeneous assembly they have a number of characteristics in common. There is first their size which, although it varies within wide limits, puts all plant viruses and the majority of animal viruses beyond the resolution of the optical microscope. Another characteristic, which all viruses, including the rickettsiae, have in common is their extremely close affinity with the living cell, outside of which they cannot multiply. No virus has ever been cultivated on a cell-free medium. Their chemical constitution is simple, and in the very small viruses consists only of protein and nucleic acid. In some viruses, at least, the nucleic acid alone is the infective unit, the protein being probably a protective covering. Many of the small viruses behave like chemicals and can be crystallized.

Viruses are outstanding in their relationship with insects and other organisms on which they rely for their transport from host to host. This relationship is an interesting one and is discussed at greater length elsewhere in the book. Finally, we may give here a definition of a virus; there are many such but the following suggested by Lwoff seems appropriate: 'Viruses are infectious, potentially pathogenic nucleoprotein entities, with only one type of nucleic acid, which reproduce from their genetic material, *are unable to grow and divide*, and are devoid of enzymes.'

Before concluding this introductory chapter it may be of interest to list some of the more important discoveries which have been made during the intensive study of viruses carried out during the last two decades, most of which are discussed in this book.

Biochemical studies have shown the important fact that the

ribonucleic acid of some viruses is alone capable of virus replication, and much new knowledge is being obtained on the virus nucleic acids and the general chemical make-up of viruses. Development of the electron microscope and the negative staining technique have allowed the ultrastructure of many viruses to be resolved, and these have added confirmation to the results previously obtained by X-ray diffraction studies. Starting with the bacterial viruses the methods of virus multiplication are gradually being elucidated. Great progress has been made in the tissue culture of viruses, and some of this is due to the use of antibiotics to keep down extraneous contamination; arising out of this work are great possibilities in the production of new vaccines.

Many new viruses, about seventy or more, have been described from the respiratory tract and alimentary canal of man and the higher animals, and new types of viruses attacking insects and other arthropods have been isolated. Several animal viruses have been obtained in crystalline form, including one from insects which occurs in iridescent crystals.

Much intensive work has also been carried out on the study of virus genetics in the bacterial viruses.

As regards the plant viruses there is new information on soil-borne viruses and on unusual types of vectors.

All these advances in knowledge have resulted in the emergence of a new science, to which the ugly but convenient name of 'virology' has been applied.

2

SOME VIRUS DISEASES

Of Man and the Higher Animals. Birds. Plants. Arthropods. Protozoa and Bacteria

It is no part of the aim of this little book to give clinical descriptions of virus diseases; it may, however, be of interest to mention some of the many diseases of this nature affecting different types of organisms and to give some of the important facts.

Virus diseases of man and the higher animals

Measles

This was first recognized as an independent disease in the seventeenth century by the English physician Sydenham, and its virus nature was established about fifty years ago. Measles is extremely prevalent, and the great majority of adults in civilized lands have suffered from it.

Great epidemics of measles have occurred from time to time in all parts of the world, such as the 'black measles' of the eighteenth century in London. Where there is 'virgin soil', that is large populations which, for some reason or other, have avoided infection, the virus may spread with terrifying speed. Examples of this are the epidemic in the Fårön Islands in 1846 and the outbreak in Greenland in 1951. Waves of measles run through populations every few years: in England, the greatest incidence of the disease is between November and March, but sporadic cases are always to be found during the off-season.

Although not generally a killing disease, it can be serious in cases of very young or elderly persons and may leave behind it much chronic disability.

Although so many new viruses affecting man are being discovered, it sometimes happens that relationships are discovered between viruses previously thought to be quite distinct. This is the case with the viruses of measles and dog distemper. Recent work has shown that the two are serologically related; thus, when the measles virus is grown in tissue culture (see chapter 5) the damage done to the cells is prevented by the addition of an antiserum prepared against dog distemper virus. Conversely, an antiserum prepared in ferrets against the measles virus neutralized the effects of distemper virus in mice.

Influenza

This famous virus disease has been known for centuries, though its virus nature was not finally determined until 1933 by Smith, Andrewes and Laidlaw. From time to time great pandemics of influenza have occurred since about the beginning of the twelfth century; the most important being after the First World War in the years 1918–19. On this occasion, about half the population of the world was attacked, and it is estimated that over 20 million people died of the disease and its complications.

An outstanding fact about the influenza virus is its variability, and the sudden appearance of a new variant probably plays a major part in starting a pandemic. In 1940 a second influenza virus was recovered in New York independently by Francis and Magill; this was called *B* to distinguish it from the earlier virus *A*. Then in 1957 a new variant of virus *A* appeared in China and spread throughout the whole world; this is known as the Asian virus.

The ever-present menace of the influenza pandemics has stimulated international collaboration, and in 1947 the World Health Organization established reference laboratories in many parts of the world. Stuart Harris describes how laboratories studying the pandemic of the Asian virus in many countries found viruses identical serologically with each other yet serologically distinct from any of the viruses recovered before this outbreak. Some interesting information on the international spread of the Asia strain of influenza has been compiled by Jensen and his colleagues from the records of the U.S. Public Health Service. This virus strain affected virtually every major country in the world within six months after the disease spread from the mainland of China to Hong Kong. The spread of the virus through Europe and North America during the summer months set the stage for epidemic disease in the autumn.

If we examine the duration of the period of time necessary for epidemic spread of influenza in three great pandemics it becomes clear how greatly the increasing mobility of man increases the mobility of the virus. Thus eleven months or more were necessary for world-wide spread of the pandemic of 1889–90, whilst only six months were sufficient for a similar degree of spread of the Asian virus. The 1918–19 pandemics fall somewhere in between the other two pandemics. Dozens of outbreaks in areas previously unaffected by the disease during the early weeks of the pandemic could be directly traced to the crews of air or surface vessels recently arrived from epidemic regions.

About 80 millions in the U.S.A. became ill during the epidemic and this represents a 50 per cent attack rate.

The common cold

Numerous viruses may be responsible for minor respiratory ailments but probably none is the cause of the common cold which many people have twice a year or more. The incubation period is normally two or three days. This is the opinion of C. H. Andrewes of the Common Cold Research Unit of the Medical Research Council at Salisbury. Research work here has been carried on for thirteen years on this very difficult and frustrating problem. Owing to the lack of an experimental animal all the work has had to be carried out on human volunteers, which is a cumbersome and time-consuming process. Much research has been carried out in attempts to propagate the virus in tissue culture. In 1953 Andrewes and his colleagues reported that a common cold virus had been propagated through ten cultures in series of human embryonic lung, colds being reproduced by materials from the fourth, sixth, seventh, ninth and tenth cultures. After that cultures produced no colds and it may be assumed that either the virus was then lost or had become too attenuated to produce colds. Now, however, in January 1960, it was announced that success had at last been achieved; agents have been obtained from three individuals with 'wild' colds and have been propagated through eight tissue-culture passages. Furthermore the virus continued to produce colds in volunteers after this continued tissue culture. In the new technique embryonic human kidney-tissue was used for the culture, and it was kept at a temperature of 33° C.; this is a rather lower temperature than is usual for tissue culture, and in addition the culture-medium was made slightly more acid than is customary. A few days after inoculation with the cold virus the kidney-cells begin to degenerate and a pathological effect was also produced by the virus in the kidney-cells of monkeys. This cytopathic effect on the cells is important because it can be used to show the presence of the cold virus without the necessity for inoculation to human volunteers. Another interesting fact is that the cold virus interferes with the multiplication of other viruses in tissue culture and this phenomenon can also be used as a means of identification.

From the work at Salisbury it seems clear that chilling alone does not induce colds, but possibly changes in humidity or other seasonal factors increase susceptibility to cold viruses, either those being carried by an individual or those transferred to him from someone else.

Poliomyelitis

This virus disease, which can be likened to a modern plague, is as widely spread as the influenza viruses and, like influenza, occurs in epidemics. The virus affects the alimentary canal and the spinal cord; it is in the latter region where the importance of the disease lies because of the resulting paralysis.

So widespread is the poliomyelitis infection that a very high proportion of the world's population has been affected at one time or another. However, many of these infections have been very mild or almost symptomless and in consequence there is a high proportion of immune persons among older people. In an epidemic therefore it is mainly children and young persons who are liable to be infected, hence the term for the disease 'infantile paralysis'.

For many years slow progress was made in the study of the disease because for long the monkey was the only experimental animal that was known to be susceptible to the virus other than man himself, and knowledge was only obtained laboriously by the sacrifice of hundreds of monkeys. However, in 1949 a great step forward was made by three American workers, Enders, Robbins and Weller, who succeeded in propagating the virus in cultures of tissues other than those of the nervous system. There was thus made available a ready source of poliovirus, which paved the way to the production of a vaccine. The first of these was the formalin-treated vaccine of Salk in America, which has now had extensive trials and has proved efficacious in reducing the number of paralytic cases of poliomyelitis. It is possible that a vaccine prepared from an active but attenuated virus, such as are used against smallpox and yellow fever, might be superior to the inactivated formolized virus, but there are obvious dangers here which require further investigation.

Even with the formolized virus there are pitfalls to be avoided, for example there is the likelihood of aggregation of virus particles so that those particles in the centre of the aggregate would be protected from the formalin. To overcome this the virus preparations are filtered through Seitz pads; this separates out the virus particles. Another pitfall is the possibility of the occurrence of a latent monkey-virus in the kidney-cells from that animal which are used in the propagation of the virus.

With the advent of the successful tissue culture, two American workers, Schaffer and Schwerdt, have been able to visualize the poliomyelitis virus on the electron microscope and even to obtain it in crystalline form. This aspect is considered further in chapter 3.

Chicken-pox (*Varicella*) and shingles (*Zoster*)

Chicken-pox is a mild though highly contagious virus disease which is to be met with practically everywhere, and though all ages are liable to attack the most susceptible age is about five or six. It is reckoned that about 52 per cent of adults have had the disease in childhood. Shingles, on the other hand, affects quite a different age-group, rather more than half of whom are 45 or over.

Much interest has been aroused by these two diseases, and there has been a good deal of speculation as to whether they were caused by the same or related viruses. At one time it was thought that a person who had suffered from chicken-pox in childhood was immune to shingles in later life. This is now known to be untrue, and the relationship between the two is much clearer.[1]

[1] For much of the information that follows I am indebted to the writing of A. W. Downie.

The idea that chicken-pox and shingles might be due to the same virus was first suggested by von Bokay in 1892. This observation has now been confirmed and it has been shown that chicken-pox acquired from contact with shingles differs in no way from chicken-pox acquired from chicken-pox. Furthermore, fluid from the vesicles of shingles can produce typical chicken-pox in susceptible children. It is now considered that shingles is caused by a clinical infection with chicken-pox in a partially immune person, that is one who has had an attack of chicken-pox in childhood. This is supported by the localized nature of shingles, its occurrence in an older age-group and a previous history of chicken-pox in many cases.

However, there is still much that is obscure about the disease of shingles. For example, cases often occur without any apparent contact with other persons suffering from shingles or chicken-pox and there is no apparent increase in the prevalence of shingles during a chicken-pox epidemic. Two suggestions have been made. One is that shingles is a localized infection in a partially immune individual acquired by contact with a healthy carrier of chicken-pox. The other, which seems more likely, is that shingles is the result of the reactivation of latent chicken-pox virus harboured by the individual for many years. This is somewhat similar to the reactivation of the *herpes simplex* virus, except that this virus may be stimulated frequently into activity, whereas with shingles a second attack is uncommon.

However, it seems likely that the recent successful propagation of the shingles–chicken-pox virus in tissue culture will help to solve some of these problems.

Rabbit myxomatosis

This now famous virus disease is of interest from two points of view: first its use in an experiment on a large scale on the biological control of an important agricultural pest and, secondly, as affording the opportunity of studying the evolution of a virus disease in an animal population that was originally completely susceptible.

Some ten or more years ago the rabbit myxoma virus was comparatively unknown, but after its deliberate introduction into Australia and its more or less accidental importation into the continent of Europe and into the British Isles it has been constantly in the public eye. In the British Isles the disease became a subject of controversy between the practical agriculturalist, who looked only for the eradication of a tiresome pest, and those who regarded the disease from a more sentimental aspect.

The disease has been studied by Fenner and his colleagues and much of our knowledge is derived from them. Myxomatosis was first recognized in 1896 in Uruguay in a laboratory rabbit colony in Montevideo. At intervals since then outbreaks have occurred in domestic European rabbits in various parts of South, Central, and North America. Before its importation into Australia and Europe the virus was unknown outside the American continent.

The natural host for the virus in South America was shown by Aragâo in 1943 to be the common wild rabbit of Brazil (*Sylvilagus brasiliensis*), in which the only manifestation of disease was a single localized tumour under the skin. Aragâo also demonstrated that the virus was transmitted mechanically from these tumours by mosquitoes. The natural host of the virus in Central and North America, where *S. brasiliensis* does not occur, is not known.

The host-range appears to be limited and the only animal other than rabbits found naturally infected in a few cases is the European hare (*Lepus europaeus*). In tissue culture, however, the virus has been induced to multiply in cells from a number of different animals.

Compared with the disease produced in the Brazilian rabbit, the effects of the virus in the European rabbit (*Oryctolagus cuniculus*) are very different. There is a very generalized infection, with great oedema of the head and genital region together with widespread tumours all over the body. The disease is a distressing one, and with the highly virulent strain of the virus the mortality rate is as high as 99 per cent.

In Australia the virus is transmitted from diseased to healthy rabbits by mosquitoes; this is a purely mechanical process and has been called by Fenner the 'Flying Pin'. In Europe the mosquito is not important and the chief vector is the rabbit flea (*Spilopsyllus cuniculi*), which again is a purely mechanical vector.

In Australia Fenner and his co-workers have demonstrated that changes in the resistance of wild rabbits to the myxoma virus have developed together with changes in the virulence of the myxoma virus itself. The extremely lethal nature of the virus makes it a very efficient selective mechanism for genetic resistance. Animals which recover from the disease do so in virtue of their genetic resistance, which is passed on to their progeny.

Like so many other viruses that of myxomatosis has undergone a number of changes, so that the original strain giving 99 per cent mortality has been replaced by attenuated (less virulent) strains; these are spread more easily by mosquitoes because of the persistent lesions produced in the skin.

There is some question as to whether myxomatosis is evolving along the same lines in Europe. H. V. Thompson considers that this is not necessarily the case as far as Britain is concerned. The principal vector is the rabbit flea, and widespread annual epizootics do not occur; attenuated and virulent strains of the virus exist side-by-side, and an increase in innate resistance to myxoma virus has not yet been demonstrated.

The myxoma virus is rather large and is included in the group of pox viruses. On the electron microscope it is indistinguishable from vaccinia virus.

Virus diseases of birds

There are many viruses which affect birds, some of them being of great economic importance. Perhaps the most famous and the most interesting from a scientific point of view is the Rous Sarcoma, so called after its discoverer Peyton Rous, who first described it in 1911. Previous to that it had been demonstrated in 1908 by Ellermann and Bang that fowl leukaemia could be transmitted by means of cell-free filtrates. These two cases are of particular interest as being examples of virus-induced cancers; this aspect is further dealt with in chapter 10.

Another virus affecting birds which is frequently in the news because of the financial losses caused is that known as fowl-pest or fowl-plague. The year 1959 was the worst yet experienced and the only means at present known to restrict its spread is the slaughter of infected birds and their contacts, a similar policy to that in practice in Great Britain with foot-and-mouth disease of cattle. Up to

November 1959 there had been 856 outbreaks and more than 1,500,000 birds had been slaughtered at a cost in compensation in excess of £1,250,000.

Psittacosis, parrot fever

This disease is of interest for two reasons, first because of its ability to infect man and secondly because of its very wide host-range. In size the virus is a large one and is placed in the same category as those causing such diseases in man as trachoma, inclusion conjunctivitis and lymphogranuloma venereum. The causal agents of these diseases all pass through a sequence of developmental forms and perhaps for this reason are regarded by some virologists as belonging to the Rickettsiae, a group of agents midway between the bacteria and the true viruses.

The disease produced in man may be only a mild fever or in more severe cases a type of pneumonia. Large doses of antibiotics are sometimes effective, which is not the case with diseases caused by the small viruses.

According to Bedson the disease was first noticed in 1880 by Ritter in Switzerland where seven cases, three of them fatal, occurred in a household which had a sick parrot. In the next twenty years outbreaks occurred in Germany, Switzerland, Italy and France. Later there was an extensive outbreak involving twelve countries including Britain; this pandemic apparently had its origin in infected parrots imported from South America.

As previously mentioned the host-range of psittacosis virus is quite extensive; natural infection occurs in many species of birds, various types of parrot, pigeons, domestic fowl, ducks, fulmar petrel, herring gulls and several species of finch. According to Meyer spontaneous infection has been recorded in at least thirty-one species belonging to nineteen genera. This question is discussed again in chapter 6 under the subject of reservoirs of virus infection.

Virus diseases of plants

We have seen in chapter 1 that the first virus discovered was one causing a disease in the tobacco plant, to which the name 'mosaic' was given. This name is now given to a large group of somewhat similar diseases, in which the chief symptom is a mottling or 'mosaic' pattern on the leaves of the affected plant. Although a large group, the mosaic-type viruses form only a part of the three hundred and more plant-viruses which are at present known. We give now short descriptions of one or two well-known mosaic diseases.

Tobacco mosaic

A tobacco plant affected with mosaic is stunted in growth, shows the typical yellow and green mottling of the leaves, and may also exhibit a certain amount of leaf-distortion depending on the strain of virus. A description of the virus itself is given elsewhere, and it will suffice here to say that it is extremely infectious to other susceptible plants and has a wide host-range affecting many different kinds of plants. Unlike the majority of plant-viruses it has no insect vector but is spread from plant to plant chiefly by man himself.

The mere act of handling infected and healthy plants is sufficient to spread the infection. One or two tomato plants infected with tomato (=tobacco) mosaic are sufficient to infect every plant in a commercial glasshouse containing several thousand, the virus being carried from plant to plant by the ordinary processes of cultivation.

The virus is extremely stable and for that reason retains its infectivity in most brands of smoking tobacco, whence it is easily spread by smokers to susceptible plants. It is thus a prolific source of infection for tomato crops in commercial houses.

It also retains its infectivity for long periods of time, and dried tobacco leaves with mosaic are still capable of causing infection in healthy plants after many years. Caldwell relates how the leaves of tomato infected with a strain of tobacco mosaic still retained infectivity after nearly twenty-five years' storage in an envelope in a drawer.

Cucumber mosaic

In some ways the virus causing this disease is the antithesis of the foregoing although the symptoms on the tobacco plant, which it can also infect, are sometimes rather like those of tobacco mosaic and it has an even wider host-range. Cucumber mosaic virus is transmitted from plant to plant by one or more species of aphids, and it is very common in most gardens, where it may infect a wide variety of ornamental plants. Many of these plants are perennial, such as delphiniums, lupins, michaelmas daisies, dahlias, privet and *Buddleia*. In consequence it is sometimes very difficult to grow the out-door cucumber or vegetable marrow in gardens where this virus is lurking in perennial plants. The disease caused in the out-door cucumber and the marrow is very serious and the plants are frequently killed.

In contrast to the virus of tobacco mosaic that of cucumber mosaic is unstable and retains its infectivity for a very short time, about 54 hours in extracted sap, and is inactivated rapidly in desiccated leaves.

Besides being of considerable economic importance in horticulture the virus has some historical interest, as it was the first to be transmitted experimentally by means of an aphid vector. We shall discuss in chapter 8 some of the interesting relationships between insects and viruses and see how far-reaching are these relationships.

Tulip mosaic

This disease is usually known as 'tulip break' because of the characteristic colour change produced by the virus in the flower of the tulip. This change or 'break' is most pronounced in self-coloured varieties and causes very attractive variegations and pencillings of the flower. Historically the disease is of great interest since it is the oldest plant virus known, and examples of 'broken' tulips occur frequently in paintings of the sixteenth and seventeenth centuries.

In his charming article 'Tulipomania the Benevolent Virus' (*Perspectives in Virology*, 1959) Dubos describes how the growing of tulips became a craze in Holland and Flanders round about the middle of the sixteenth century. Commercial growers and wealthy citizens competed for the production of new colours or patterns of pigmentation, and it was here that the virus played a predominant part. Huge sums, 13,000 florins in one case, were paid for a single bulb, and a young woman became a very desirable bride when a famous bulb was made the sole item of her dowry.

Although it is true that some tulips which are infected with the virus have been in cultivation for 350 years, others have gradually deteriorated and dropped out owing to the deleterious effect of the disease. Indeed, even in the seventeenth century it was realized by some horticulturists that the tulips with 'broken' colours lagged behind in vigour, compared to tulips which were free of the infection.

Abutilon mosaic

This is another rare example of a virus which improves a plant's appearance. The variegated plant (*Abutilon* sp.) was introduced into the British Isles about 1868 and became popular as an ornamental plant. The leaves show a very striking variegation of greens and yellows as compared with the uniform green of the uninfected plant. By grafting scions of variegated plants to green shoots of normal plants it was discovered that this variegation was infectious. Since, however, this condition was never known to spread from variegated to non-variegated plants, and since, until lately, no alternative method for making it spread, other than grafting, was known, it was called an 'infectious variegation' and thought to be in a different category from other mosaic diseases. However, it has now been shown by workers in South America that the virus is transmitted by a specific insect vector, a species of whitefly, which does not occur in the British Isles. The reason for the lack of spread of *Abutilon* mosaic thus becomes clear.

Virus diseases of arthropods

Until recently no viruses were known to affect any animals outside the vertebrates except the insects, but now research has discovered viruses in those minute plant parasites known as mites which, while belonging to the arthropods, are not insects. Moreover, going still further afield, there is now evidence of a virus infection of nematode worms; this is not the same thing as a nematode worm acting as the vector of a plant virus, which we shall discuss later.

Silkworm jaundice

A detailed account is given in chapter 8 of the viruses which attack insects, so for the present we are only concerned with a brief description of a representative virus disease of insects for comparison with the virus diseases of other organisms.

'Jaundice' of the silkworm has been known since 1527 when it was described by Vida in his poem *De bombicum*. The name arises from the jaundiced or yellow appearance of the affected caterpillars; the disease is very infectious and is spread by the contamination of the food-plant.

Jaundice is one of the polyhedral diseases of the nuclear type in which the rod-shaped virus particles multiply in the cell nuclei of the fat, blood, skin and tracheae. A more detailed description of this kind of insect virus is given in chapter 8; for the present it is sufficient to say that the virus particles are enclosed within protein crystals, the *polyhedra*. When the caterpillar dies, the skin which has been attacked breaks and liberates vast quantities of the virus-containing polyhedra. These serve as a most efficient mode of dissemination of the virus since they are resistant to temperature and humidity and are easily spread about. They contaminate the food-plant and are eaten by other healthy caterpillars. This is the main mode of spread of the virus, though in the case of some other insect viruses infection is undoubtedly passed from parent to offspring.

For many years before the virus origin of jaundice was discovered there was much controversy as to the nature of these peculiar polyhedral bodies. Various suggestions were put forward, the causative agent was a bacterium, a sporozoon or a chlamydozoon. All these ideas were dropped and attention was centred on a virus as the cause, and at first the polyhedra were considered to be crystalline accumulations of virus. However, various workers, notably Komarek and Breindl, noticed many minute particles inside the polyhedra which they believed might be the infectious agent, and this was finally proved by Bergold in 1948 who dissolved the polyhedra in weak alkali and observed the liberated virus-rods under the electron microscope.

Jaundice is of considerable economic importance in the silkworm industry because it is very difficult to control, since it is liable to be latent in some caterpillars. These latent virus infections sometimes become suddenly virulent and start an epizootic among the silkworm population.

Acarina: mites

Many of these minute eight-legged creatures are serious pests of plants causing much injury by feeding in vast numbers on the leaves. In addition at least four virus diseases affecting important crops are transmitted by mites.

However, it is only just recently that a virus has been found which actually attacks and kills some of these mites. The species affected are known as 'red spiders' or 'spider mites' and they are among some of the worst pests of plants. In the British Isles there is a species which attacks glasshouse crops, especially tomato plants, and another is known as the fruit-tree red spider. In California there is the citrus red mite, *Panonychus citri* Meg., and it is this species in which a virus has been discovered; not very much is known as yet about the virus except that it is very small, about the size of some of the smaller plant viruses, and causes in the red mite an infectious and very fatal disease.

Virus diseases of protozoa and bacteria

Protozoa

***Paramecium aurelia*, kappa-particles.** *Paramecium* is a minute free-swimming organism sometimes known as the 'slipper animalcule'. It measures about 100–150μ in length and swims by means of thousands of cilia; these are arranged in parallel rows and beat rhythmically, propelling the animal in a characteristic manner. The cilia also procure the food by directing a current which carries bacteria into the mouth.

For some years a peculiar state of affairs has been observed in *Paramecium*, in which certain strains contain large numbers of particles known as 'kappa-particles'. Such individuals are known as *killers* because they liberate into the culture-medium something which kills other strains of *Paramecium* known as *sensitives* and which do not contain the kappa-particles.

In the earlier studies of the kappa-particles they were thought to be natural genetic elements of the cytoplasm and were interpreted as plasmagenes. Later work, however, proved that the particles were infective and contained a considerable quantity of DNA (see p. 37). Various suggestions were then put forward as to the nature of kappa; some authors considered it was a virus, others a rickettsia, a bacterium or even a degenerative type of alga. Of these possibilities it seems more likely that the kappa is a virus since the DNA is distributed throughout the particle; there is no cytoplasmic region and apparently no enzymes are present. On the other hand there are some affinities to bacteria in size, reproduction by division and susceptibility to antibiotics. Sonneborn (1959), who has made a comprehensive review of the whole subject, suggests that, since kappa has affinities with both viruses and bacteria, it may be a kind of 'missing link' between the two.

Bacteria

It was in 1917 that Twort, an English bacteriologist, was examining some bacterial cultures on a plate of agar or a similar medium when he noticed a number of clear areas in the colonies where the bacteria had apparently been destroyed. Twort, like Alexander Fleming some five years later when he discovered penicillin, was curious as to the meaning of these clear patches, and, instead of throwing away the culture, he passed it through a bacteria-proof filter-candle and seeded the filtrate into a fresh culture of the bacterium. He found to his surprise that more clear areas appeared in the culture, and realized that a filter-passing agent which had the power to multiply, in other words a virus, was concerned. A little later the same phenomenon was demonstrated independently by d'Herelle, a Canadian worker, who coined the phrase 'bacteriophage', or 'bacterium eater'. Nowadays we speak usually of bacterial viruses or occasionally 'phages.

At the time of the discovery of the bacterial viruses great hopes were held that the key to the conquest of bacterial diseases had been found, and Sinclair Lewis wrote his book *Arrowsmith* around this theme. Unfortunately these hopes were not realized, but Fleming's discovery of penicillin a few years later achieved everything that had been forecast for the bacteriophages.

The bacterial viruses have been subjected to perhaps more intensive study than any other virus, and being a single cell the bacterium is ideally suited for this. The mechanism of reproduction has been exactly worked out, the morphology of the phage-particles is accurately known and the ultrastructure of the component parts has been characterized with great beauty on the electron microscope. In addition the genetic analysis of phage reproduction has been intensively studied; there is now a vast literature on the bacterial viruses, and for further information on the subject the reader should refer to volume 2 of *The Viruses* (New York and London: Academic Press).

3

HOW THE VIRUSES THEMSELVES ARE STUDIED

Isolation of Viruses. Chemical Nature. Virus crystals

Isolation of viruses

Before a virus can be visualized on the electron microscope, before its physical properties and chemical nature can be examined and before one can be certain that a virus is involved at all in a particular study, it must be separated from the constituents of the host. This is what the term 'isolation of viruses' implies, and it covers a wide range of procedures and techniques.

Quite a number of the advances in our fundamental knowledge of viruses in general have been made during studies of the plant-viruses, and the first isolation of a virus in appreciable quantities was made in 1935 by an American biochemist, W. M. Stanley, with the virus of tobacco mosaic. This achievement and the pioneer work of Bawden and Pirie have removed much of the aura of mystery which for many years hung around and impeded the study of viruses. For the first time it was possible to visualize a virus as a definite entity and not simply as a disease syndrome for which no disease agent was visible or obtainable. Since those early days the investigations on the viruses themselves as distinct from virus diseases have advanced to such a degree that Dr Roy Markham can speak of the plant-viruses as 'commonplace chemicals used for calibrating physical apparatus of various kinds and used as sources of nucleic acids'. This is true enough, though commonplace is hardly the word to describe chemicals endowed with the power of multiplication and mutation and holding the key perhaps to much greater advances in our knowledge of the living cell, and the way living things are made.

Plant viruses

The plant virologist has two great advantages over his colleague working with animal viruses: much greater quantities of virus are available and they are easier to extract. Indeed it is really rather surprising that the virus of tobacco mosaic was not isolated long before 1935; probably the common belief that viruses were merely very small bacteria discouraged a chemical approach to the subject.

Before any isolation-process can be started, it is necessary to have a large supply of virus, and that implies growing many infected plants. With some viruses this is a fairly straightforward procedure, though there are one or two points to remember. The plants should be inoculated when young, are best grown from seed, and must be grown in an insect-proof glasshouse to avoid contamination with insect-borne viruses. Plants which have much dark pigment or tannins in their sap are unsuitable, as are plants which have mucilaginous sap.

The tobacco plant is a very suitable host for the mass production of many viruses, its chief drawback being its liability to contamination with the ubiquitous tobacco mosaic virus.

There are two main methods of purification and isolation of plant-viruses, by chemical precipitation and ultracentrifugation, and in practice the best application is a combination of the two. The leaves of the infected plants are gathered as soon as the disease is fully systemic, that is, when the virus is distributed throughout the plant. (There are some viruses which do not spread beyond the inoculated leaves; in such cases only those leaves are harvested.) The next step is to extract the sap and this is best done with an ordinary domestic meat-grinder with a worm which compresses the leaves before they reach the cutters. If a quantity of leaves is being used a large screw press is convenient. Once the sap has been extracted it must be clarified by the removal of the grosser plant-material, and it is sometimes convenient here to add a reducing-agent to prevent oxidation. Preliminary clarification is best achieved by passing the sap through fine muslin and then centrifuging at about 3000 rev./min.

Sometimes it is convenient to precipitate the plant-proteins by heating the sap to 55° C for a short time and then centrifuging-off the bulky precipitate thus formed. This method is only applicable, of course, to certain viruses which are not inactivated by short exposures to temperatures around 55° C. Another method of sap-clarification can be achieved by the addition of ethyl alcohol (ethanol) at the rate of 300 ml. of 90 per cent ethanol to each titre of strained sap. There is, however, the danger that some viruses might also be precipitated by this method.

The next procedure is to precipitate the virus, and this is usually done by the addition of ammonium sulphate at one-third saturation. The precipitate is centrifuged off and the supernatant fluid discarded; the precipitate is resuspended in water or buffer solution and centrifuged on the ultracentrifuge. The procedure is repeated once or twice, the precipitate being resuspended in about one-tenth to one-quarter of the volume of the original sap. The final product is the purified virus. This, of course, is a very brief outline of the chemical-precipitation method and there are many refinements of technique that the reader can look up for himself in the literature cited at the end of the book.

Some plant viruses are not amenable to chemical methods of precipitation because of their instability, and in such cases purification can be achieved by means of ultracentrifugation only of the clarified sap. The pellet of virus obtained is then suspended in buffer and subjected to a cycle of low- and high-speed centrifugations.

The final sample of a purified virus should be free of pigment and other impurities; when suspended in water or buffer it usually has an opalescence characteristic of virus-suspensions. In the case of the rod-shaped viruses, however, particularly that of tobacco mosaic, the final preparation has a silken sheen which is quite unmistakable. When viewed by polarized light the suspension shows the phenomenon of double refraction or anisotropy of flow.

Animal viruses

The purification of animal viruses is a more difficult and tedious process than that of isolating plant-viruses; the procedure of pounding up the whole plant and extracting the juice is obviously not applicable to animals! Nevertheless, many of the same techniques together with others are used once the normal crude macromolecular constituents of the cell have been eliminated. These methods include ultracentrifugation, adsorption and elution, precipitation by salts or alcohol, and digestion of protein with enzymes. In some cases small laboratory animals are used for the propagation of the virus, and these are perhaps partly analogous to the tobacco plant of the plant virologist. One method of virus propagation, however, is used by animal virologists which is not yet, unfortunately, applicable to plant viruses, and that is the technique of tissue culture.

For purifying influenza virus the method of propagating it in the chorioallantoic fluid of 10–13-day-old chick embryos is often used. The virus attains a high concentration in this fluid, which can be made the starting-point for most purification techniques.

A procedure for the preliminary purification and concentration of influenza virus is to adsorb the virus on to chicken red-blood cells in the cold. It can then be eluted at 37° C. and in one-tenth the original volume; next it is concentrated by ultracentrifugation.

The influenza virus is comparatively large, but the isolation of very small animal-viruses is more difficult because of the presence of normal particles of about the same size. Recently Weil and his co-workers have succeeded in purifying the virus of mouse encephalomyocarditis, which measures only 30 mμ. The starting-material was mouse-brain suspensions in distilled water; and the purification was achieved by centrifugation and by the use of trypsin.

Schaffer and Schwerdt have studied the purification and properties of the virus of poliomyelitis; and for this they used virus from tissue culture. This was precipitated from the tissue-culture fluid with 15 per cent methyl alcohol (methanol), and subjected to a cycle of ultracentrifugations. Finally, it was centrifuged in a solution of sucrose; this is known as a sucrose density-gradient which separates out the various components in a solution; these appear as bands across the fluid in the centrifuge tube. In the case of the poliovirus there were four visible components when the virus was obtained from Maitland-type tissue cultures (minced tissue in a nutrient-medium) and two when the virus was obtained from monolayer tissue-culture fluids. The infectivity was associated with the lowest band, which contained only a homogeneous population of spherical particles.

These examples of the purification of the viruses of the higher animals will suffice to illustrate the general principles of the methods used.

As a contrast we give a short description of the isolation of some viruses from insects and other arthropods which are, in spite of the opinion of some animal virologists, also animals.

As we shall see in chapter 8, some large groups of insect viruses are unique in the sense that the virus particles, which are rod-shaped, are enclosed in protein crystals known as 'polyhedra'. In this type of insect-virus disease, therefore, the first step is to isolate the virus-containing crystals; isolation of the virus itself comes next. As compared with the purification procedures we have just discussed the isolation of this kind of insect virus seems comparatively simple.

A large quantity of larvae, depending on the numbers available, which have died from either a polyhedral or granular disease, are suspended in water contained in a conical flask. They are then left for a period until the bodies are disintegrated and the polyhedral or granular bodies have sedimented to the bottom of the flask; the granules, being smaller than the polyhedra, take considerably longer to deposit. The supernatant fluid containing the debris of the larvae is decanted, and the sedimented crystals are subjected to a series of centrifuge spins until an almost white suspension is obtained. A better method, perhaps, is not to wait for bacterial putrefaction to liberate the polyhedra, since this is liable to damage the polyhedra and their occluded virus, but to grind up the bodies in a blender and centrifuge off the polyhedra or granules. The next step is to liberate the virus particles from the occluding polyhedra or granules and this is done by dissolving them in weak alkali. It has been shown by Bergold that the addition of the alkali is rather critical, since if too much sodium carbonate is added the polyhedra or granules dissolve too fast and the virus-rods are dissolved as well. The next step is to spin off on the centrifuge the undissolved fragments of polyhedra or granules and other debris; this is done by centrifuging for about 5 min. at 6000 rev./min. The transparent slightly turbid suspension is then centrifuged for 1 hr at 12,000 rev./min; the supernatant fluid is poured off and the bluish-white pellet consists of the virus.

Perhaps the easiest of all viruses to purify is one attacking the larva of the common crane-fly or daddy-long-legs, *Tipula paludosa*, and known as the *Tipula* iridescent virus (see chapter 8). This occurs in very high concentration in the host and, indeed, represents 25 per cent of the dry weight of the insect. It occurs freely in the tissues without polyhedra or granular inclusion bodies. No chemical precipitation methods are needed to isolate this virus, and centrifugation only is required. The dead infected larvae are stored in water in a conical flask, similarly to the polyhedral viruses, and allowed to decompose. The supernatant fluid and the debris are decanted after standing for a week or two, and the sedimented virus is purified by a cycle of high- and slow-speed centrifugations. The final result is a highly iridescent pellet of virus in the form of large numbers of microcrystals. As the time for the virus to sediment from the decaying larvae is rather long the process can be speeded up by putting the infected larvae in a homogenizer, or simply grinding them up in a pestle and mortar. The debris is then carefully filtered off on fine muslin, and the filtrate subjected to the usual centrifugation. The mass production of this virus has now been developed. It has been found possible to infect the caterpillars of the large white butterfly, *Pieris brassicae*, with the *Tipula* iridescent virus, and these caterpillars can be raised in large numbers right through the year. By this means it is easy to produce a gram of pure virus.

To complete this short account of the isolation and purification of viruses it may be of interest to describe the isolation of a virus from a type of organism from which no virus has previously been recorded. This is a minute pest of plants known as a red spider or spider-mite and belongs to the order Arachnida in the Arthropoda. We shall read of species of mites acting as vectors of viruses in chapter 7, but this is the first mention of a virus actually attacking the mite itself.

Isolation of this virus is again fairly simple: 25 mg of infected mites are ground up in a mortar and suspended in 50 ml. of distilled water; they are then subjected to a further comminution in a glass homogenizer. The fluid is then given a slow-speed spin to remove the debris, followed by 40 min at high speed, 40,000 rev./min on the Spinco ultracentrifuge. One or two cycles of low- and high-speed centrifuging yield a pellet which contains the virus; the quantity of course is very small in view of the size of the organism.

Chemical nature

Shortly after Stanley isolated the virus of tobacco mosaic in 1935, Bawden and Pirie and their co-workers showed that it was a nucleoprotein, and the same is true of all viruses so far isolated and purified. The nucleic-acid component of plant viruses is ribonucleic acid (RNA), and in animal viruses RNA or deoxyribonucleic acid (DNA), whilst in the bacterial viruses examined so far it is exclusively DNA. In the very small viruses there is no other constituent except protein and nucleic acid, though in some of the larger viruses from the higher animals, such as vaccinia, lipids are also present.

The tobacco mosaic virus consists almost entirely of protein and ribonucleic acid in the proportion 94·4:5·6. The general elementary composition is approximately C, 50 per cent; H, 7 per cent; N, 16·7 per cent; S, 0·2 per cent; and P, 0·54 per cent. Variation in these values is reflected in the minor amino-acid differences found in various strains. Each protein subunit in the tobacco mosaic virus (TMV) particle is a single peptide chain containing about 145 amino-acids. The carboxyl end of the chain is terminated by threonine, the sequence being proline, alanine and threonine.

New information bearing on the structure of TMV, and of its protein in particular, comes from the work in 1955 of Harris and Knight who observed that the enzyme carboxypeptidase liberates one, and only one, amino-acid residue from TMV and that this amino-acid is threonine. Carboxypeptidase only attacks one end, the carboxyl-terminal end, of the protein subunits, and the action of the enzyme is stopped in this case after the first link is broken. The second one happens quite by chance to be resistant, the proline-alanine link not being broken by the enzyme. The number of threonine molecules liberated indicates that there is one for every 17,000 in molecular weight, while chemical methods of analysis give similar results. The bulk of the evidence therefore is that the tobacco mosaic virus is made up of small protein subunits of about 17,000–18,000 in molecular weight, all of which have similar, if not identical, structures; they are arranged in some regular order to form a tubular molecule containing a core of nucleic acid and are composed of about 2800 of such subunits. This chemical evidence on the structure of TMV is amply confirmed by the results of X-ray analysis (plate VII).

The chemical composition of one of the larger animal viruses, vaccinia, is protein 62 per cent; lipid 33 per cent; carbohydrate 4 per cent; RNA 0·8 per cent.

Virus crystals

Because of their small size and great uniformity, a number of viruses have been obtained in crystalline form. This is particularly true of the plant viruses, and examples are those of tomato bushy stunt, tobacco necrosis, turnip yellow mosaic, southern bean mosaic and others. The crystals of tobacco necrosis virus are very suitable for electron microscopy because of their flat shape and resistance to the electron beam. Replicas of these crystals show clearly the orientation of the virus particles in the crystalline lattice (plate IV).

After the crystallization of the plant viruses was achieved, the production of some animal viruses in the crystalline state passed without much comment. The first of these viruses to be crystallized was that of poliomyelitis and the next the *Tipula* iridescent virus (plates II and III). The crystal structure of *Tipula* iridescent virus is face-centred cubic with the particles held apart by long-range forces. These crystals show Bragg reflexions of visible light and are responsible for the blue or violet colour of infected larvae.

As we shall see in the next chapter many of the small viruses are polyhedral in shape. Electron micrographs by Steere and Schaffer of replicas of a plane surface cut through the interior of a single crystal of poliomyelitis virus show this very well.

Morgan and Rose have studied ultrathin sections of adenoviruses and a particular strain of herpes virus in tissue culture with HeLa cells. They observed the formation of intracellular crystals of both these viruses which appeared to be of a cubic body-centred lattice.

4

ELECTRON MICROSCOPY OF VIRUSES

Some Techniques of Electron Microscopy. Size and Shape of Virus Particles. Ultrastructure of Viruses. Behaviour of Viruses in the Cell

Some techniques of electron microscopy

The conception of a virus as a mysterious, almost super-natural, entity dating from Mayer's studies on tobacco mosaic virus in 1886 persisted up to quite recent times, and learned discussions were held as to whether viruses were particulate or not. No one, however, suggested what they were if they were not particulate, and Beijerinck's famous conception of a *contagium vivum fluidum* was still quoted. As lately as 1932, T. M. Rivers stated with a good deal of justification that the size of no virus was accurately known.

The first indication of the shape of a virus was given by two American workers, Takahashi and Rawlins, in 1933, who observed that the sap of a tobacco plant infected with the mosaic disease showed the phenomenon of double refraction when made to flow and viewed by polarized light. This indicates that the sap contains rod-like entities, an observation amply confirmed in later years.

Apart from the so-called elementary bodies of certain animal-virus diseases which came within the resolution of the optical microscope, the first actual picture of a virus particle was made by Barnard with his ultra-violet-light microscope. It was only possible, however, to resolve some of the larger animal-pox-viruses, and no details of particles could be seen. It was not until the invention of the electron microscope that any conception of what a virus really looked like was obtained.

The improvement in the electron microscope and the developments in the techniques for the study of viruses during the past ten years have been very great, and it is only possible to indicate them here without any technical description.

The first virus to be observed on the electron microscope was that of tobacco mosaic, and the particles were shown to be rod-shaped, as had been predicted earlier by other methods. The first electron micrographs of this virus showed only that it was a rod-shaped particle and no other details were visible. It is by comparing these early pictures with modern electron micrographs of the same virus that some idea is obtained of the great advances which have been made in electron microscopy.

Metal shadowing

In the early days the object to be examined was put straight on to the film-covered grid and examined in the microscope, but in 1946 Williams and Wyckoff developed a method of shadow-casting in which metal, evaporated from an electrically heated filament in a vacuum, impinges on a specimen from a sufficient distance so that the metal atoms arrive in almost parallel straight lines. The metal, being more opaque to the electron beam, casts a shadow, thereby giving a three-dimensional effect to the objects; this gives micrographs which are a great improvement on the flat and rather characterless pictures previously obtained. Metals used for shadow-casting should not show a granularity which might obscure the structural detail of the object; several have been used, and they include chromium, gold and platinum alloys, and uranium.

Thin sections

The development of the ultra-microtome for cutting ultrathin sections has marked a great advance in electron microscopy. In the early days sections were cut at a high speed and of a wedge shape, the thin end only being examined. Nowadays, however, the technique has become almost ordinary procedure.

There are various types of ultra-microtomes on the market; one or two still make use of steel knives, but the glass knife, made by fracturing a strip of plate glass, is more widely used. Recently, the diamond knife, made from a polished industrial diamond, has become popular. The advantages of this are its extreme hardness, its durability—it can be used continuously instead of only once as with the glass knife—and its suitability for cutting hard objects. Very fine results have been achieved with the diamond knife but it requires great care in its manipulation (plate XII).

A good operator with either glass or diamond knives should be able to cut sections of 300–400 Å in thickness.

The embedding medium is important, the most popular being a plastic made by the polymerization of butyl and methyl methacrylate. Other media are 'Vestopal' polyester resin and 'Araldite' expoxy-resin; all these require dehydration with ethanol. Just recently, however, another embedding medium, to which the name 'Aquon' is given, has been described; this medium is miscible in water and avoids the necessity of using dehydrating agents.

New staining methods

The standard method of fixing virus materials for electron microscopy is to use buffered 1 per cent osmium tetroxide, and this acts as a stain as well as a fixative. Recently, however, a number of new staining techniques have been developed (these are not really stains, *sensu stricto*, but serve to enhance the contrast by an increase in electron density); one of the most successful of these is phosphotungstic acid, and Selby has used this acid at a concentration of 1 per cent adjusted to pH 5·4. A technique of 'negative staining' developed by Brenner and Horne in Cambridge has proved a great advance; this consists in spraying the virus particles on the grid with 2 per cent phosphotungstic acid adjusted to pH 7·0 with potassium hydroxide. As we shall see later this technique enables the ultrastructure of the virus particles to be visualized.

Other useful stains are 1 per cent phosphomolybdic acid buffered at pH 5·4 with phthalate buffer, chrome gallocyanin, 0·1 M ferric chloride, mercuric bromophenol blue and 0·5 per cent uranium acetate buffered at pH 6·1 with veronal acetate buffer. The last named has proved very successful with insect-virus sections.

Supporting films

Before the specimen can be examined it has to be placed on a very thin film, which in turn is placed on the copper grid for insertion in the electron microscope. For several years the standard supporting-film on the grid was one of collodion or 'Formvar', and these are still used for routine work. The increasing resolution of the electron microscope and the decreasing thickness of the sections, however, led to difficulties and it became necessary to find another substrate. Carbon films made by boiling carbon in a vacuum have proved to be very satisfactory; they have virtually no structure to confuse the issue at high magnifications, they are more resistant to the electron beam and they do not volatilize and foul the gun.

Carbon replicas

It is sometimes useful to be able to examine the surface of a specimen by means of a replica, and this is important when dealing with virus crystals. Several replica techniques have been devised since the electron microscope became a laboratory tool, but except for specialized work the carbon replica has proved the most successful. This technique was first suggested by Bradley, working at the A.E.I. Research Laboratory at Aldermaston. The carbon replica technique is a single-stage process, in which the suspension of material to be replicated is dried down on to a 'Formvar'-coated grid. This is coated with carbon from a source normal to the grid surface and the material removed. The source of carbon consists of two pointed carbons, spring-loaded and touching at the points. A heavy current is then passed across and a spot temperature of 2000° C. is reached; the carbon boils, and a thin coating of structureless carbon is deposited on the grid. The 'Formvar' and material are then removed by their respective solvents (plate III, B). This technique has been applied in various fields and a resolution of 7 Å has been obtained.

Freeze drying

The most serious source of structural deformation in biological specimens is the force of surface-tension, and attempts have been made to adapt the old-established technique of freeze drying to electron microscopy in order to avoid this type of distortion. Two methods are current, one devised by Anderson and the other by Williams.

Anderson's technique avoids surface-tension forces by drying out in liquid carbon dioxide, but Williams's method is more suited for general work. A suspension of the liquid is sprayed on to a carbon-coated copper disk held at the temperature of liquid nitrogen. The disk is put in a vacuum, the temperature raised to -50° C, and the ice allowed to sublime. The carbon film is removed and placed on specimen grids ready for examination.

Summary

By the application of these modern techniques to the electron microscopy of viruses, it is now possible to obtain information on the following aspects: (1) size and shape of virus particles; (2) topography of virus particles; (3) surface detail; (4) internal structure; (5) methods of replication; (6) behaviour of viruses in the cell; (7) virus crystals; (8) latent viruses.

Size and shape of virus particles

The virus of tobacco mosaic, the first virus to be discovered, the first to be isolated in large amounts, and the first to be seen on the electron microscope, was also the first to have its shape and structure definitely established. We have seen how Takahashi and Rawlings in 1935 noticed the double refraction exhibited by tobacco mosaic sap, which indicated that the virus was rod-shaped. This observation was amply confirmed later on by the electron microscope, which showed the virus to be a rigid rod measuring 350×15 mμ (1 mμ equals one millionth of a millimetre) (plate V, A).

A characteristic of the smaller viruses is their extreme uniformity of size; this expresses itself in the ability to form crystals in which all the particles are pointing the same way and are identically situated. This uniformity of size makes viruses useful for physical studies.

There are four main methods for studying virus size and structure: by chemical analysis, by hydrodynamic methods, by X-ray crystallography and by electron microscopy. It is with the last method that we are mainly concerned here. By means of the electron microscope the size and shape of virus particles are easily determined and special staining methods can delineate details of surface structure. Improvements in technique make it possible to cut ultrathin sections of virus particles and this, combined with the special staining methods, shows some of the internal structure.

There are many rod-shaped plant viruses besides that of tobacco mosaic and these include potato viruses *X* and *Y* and a number of soil-borne viruses, some of which are latent in their plant hosts.

Many plant viruses are spherical or nearly spherical and are very small; examples are the viruses of tomato bushy stunt, southern bean mosaic, turnip yellow mosaic, tobacco ringspot and squash mosaic, all of which range in size from 27 to 30 mμ. These very small viruses are extremely uniform in size and shape and so are easily obtained in crystalline form. While the majority of the viruses affecting animals and the higher plants are either rod-shaped or nearly spherical, a number of the bacterial viruses are tadpole- or sperm-shaped with a 'head' and a 'tail'.

Air-dried preparations of virus particles in the electron microscope are apt to show distortion and to appear smaller than they actually are. Accurate estimation of size, therefore, can only be made on frozen-dried specimens.

It was suggested by Crick and Watson that all small viruses are built up on a framework of identical protein subunits packed together in a regular manner. They showed that three types of symmetry are possible for spherical viruses, namely, tetrahedral, octahedral, and icosahedral.

By the use of freeze drying and by a manipulation of the metal shadowing technique it is possible to obtain evidence on the topography of a virus particle. Williams and Smith were able to demonstrate unequivocally that the particle of the *Tipula* iridescent virus is an icosahedron with twenty faces; this was done by means of double shadowing. A cardboard model of an icosahedron was made and shadowed by two light-sources separated 60° in azimuth and orientated so that an apex of the hexagonal contour points directly towards each light source. Under these conditions two shadows are thrown: one has five sides and is blunt ended and the other has four sides and is sharp ended. When a frozen-dried particle of *Tipula* iridescent virus (TIV) is treated similarly, two shadows identical with those thrown by the model are produced. By means of this shadow analysis it is proved that the *Tipula* iridescent virus particle is an icosahedron (plate VI, A and B).

Ultrastructure of viruses

Until recently the amount of external structure of virus particles which could be seen by means of the electron microscope was somewhat disappointing. It was not possible, for example, to make out the helical structure of the tobacco mosaic virus rod as revealed by X-ray studies (plate VII). The new staining methods, however, especially negative staining with phosphotungstic acid developed by Brenner and Horne in Cambridge, have given greatly improved results. By this means the arrangement of the protein subunits of virus particles can be visualized directly on the microscope.

Tobacco mosaic virus

Most of our information on the ultrastructure of TMV is derived from the earlier X-ray studies, but confirmation of this is now forthcoming from studies on the electron microscope using the new staining techniques. We know that the smaller viruses consist of two distinct and separable parts, the protein and the nucleic acid, ribonucleic acid (RNA) in the case of the plant viruses and RNA or sometimes deoxyribonucleic acid (DNA) in the case of the animal viruses. We have now to see how the protein and the RNA are put together to form the rod of TMV.

It was first suggested by Bernal and Fankuchen in 1941, on the results of their X-ray studies, that the protein of TMV was built up of subunits and this was further indicated by end-group analyses carried out by Harris and Knight and others. In 1954 Watson showed by means of X-ray diffraction that the structural subunits of the particle form a helix, and more recent work by Franklin and her co-workers, using TMV to which atoms of mercury had been attached, have suggested that there are 49 subunits on three turns of the helix. This gives a total of 2130 subunits in a complete rod, compared with the figure of 2900 repeating subunits, of molecular weight about 17,000, calculated by Knight using chemical methods.

As we shall see shortly the small plant viruses have a much smaller number of subunits, and these can be counted directly on the electron microscope using the negative staining technique (plate XI, A).

The RNA of the TMV particle is thought to be only 6 per cent of the particle and appears to be all in one piece, a single-strand polymer following the pitch of the helix. In 1955 Hart showed that it was possible to remove part of the protein coat from the TMV particle, leaving the RNA protruding and apparently intact, and further that this RNA could be covered up again when more TMV protein was supplied. This suggests that the RNA does not run through individual protein subunits, but rather that the subunits pack together in such a way as to leave space for the RNA between them. In plate VII is a model of the TMV particle based on the results of X-ray studies by Franklin and her colleagues. Some of the protein subunits have been removed to show the suspected position of the RNA in relation to them.

Tobacco rattle virus

Nixon and Harrison have recently examined another rod-shaped plant-virus, known as tobacco rattle virus and which is soil-transmitted. They have applied some of the new 'electron stains' in their study and it is interesting to consider their results in the light of what is known about the structure of the tobacco mosaic virus. They found that the particles were tubular, with a central hole of approximately 4 $m\mu$ dia., and an outside diameter which varied from 17 to 25 $m\mu$ according to the treatment. Next to the central hole, which could be filled with uranyl acetate, was a region 1–1·5 $m\mu$ thick, which stained heavily with osmium tetroxide or the other stains. The rest of the particle stained lightly and showed transverse bands 2·5 $m\mu$ apart. It is suggested that these bands may represent a helical structure similar to that of tobacco mosaic virus, which tobacco rattle virus resembles in many respects. Nixon and Harrison are not certain about the exact position of the nucleic acid within the virus particle, but they suggest that it may be located in a region lying immediately outside the part lining the central hole.

Some other plant and animal viruses

There are numerous other plant viruses which have rod-shaped particles, and one of them, the virus of sugar beet yellows, has been studied on the electron microscope by Horne, Russell and Trim at Cambridge, using the new negative staining technique. Their studies show that the beet virus resembles that of tobacco mosaic in having a central hole 30–40 Å in diameter. Although the postulated helical structure of TMV has not been observed in the beet virus, at very high magnifications individual filaments were observed to have a regular periodic structure along the axis. The mean distance measured between the small regular structures was about 26–30 Å and the width of the individual units approximately 20 Å. It is thought that the most likely interpretation of these results is a loose hollow helix.

In the case of the very small near-spherical plant viruses such as that of turnip yellow mosaic (TYM), it was shown by Markham that the RNA is contained in the centre of the particle and is surrounded by a protein coat.

In earlier electron micrographs of the spherical viruses using the conventional shadow technique, it was sometimes possible, if the resolution was good, to observe that the surface of the particle was irregular and bumpy. Now, however, with negative staining it is possible actually to count these 'knobs' on the surface or in other words, the protein subunits of the outer coat. The first pictures along these lines were made by Horne and his co-workers at the Cavendish Laboratory in Cambridge of adenovirus and herpes virus, and the nature of the arrangement of the protein subunits can easily be deduced from the micrograph reproduced in plate VIII A. The particle has a roughly hexagonal outline and is clearly faceted, each facet being an equilateral triangle with a side consisting of six subunits. The presence of triangular facets excludes a dodecahedral form and suggests that the particle is an icosahedron. A model of an icosahedron with a side of six subunits has been built from table-tennis balls and plate VIII, B is a photograph of this model standing on an edge. The arrangement of the subunits in the virus particle and in the model are identical. Herpes virus shows a similar subunit structure but the subunits appear to be hollow and ring-shaped. Hollow or tubular subunits have also been observed in the TIV particle (see also plate IX).

Klug and Caspar have stated that all the small viruses so far studied by X-ray methods appear to have icosahedral symmetry, which implies that they consist of sixty asymmetric units; these can actually be arranged in an infinite number of slightly different models, but all can be visualized as consisting of twelve identical groups of five units, equivalently related by the five-fold axes of the icosahedral point group.

The negative-staining technique has entirely changed the type of electron micrograph which can now be obtained of the small viruses, especially those from plants. It is now possible to count the protein subunits on the face of the particle, and the number of morphological units can be found by adding some combination of the numbers of 12, 20, 30 and 60 or a multiple of 60 appropriate for the particular clustering arrangement (Klug and Caspar, 1960).

In the spherical viruses it is almost possible to deduce from the picture of the subunit arrangement on the surface of the particle what is the total number of subunits by counting the number on the edges, but the knowledge of the internal detail is dependent on X-ray diffraction studies, except for what can be obtained by examination on the electron microscope of ultrathin sections of the virus particles.

Thus in the adenovirus there are six subunits on the edges giving a total of 252; there appears to be no virus described as yet with five subunits on the edge with a total of 162 subunits. Next comes tomato bushy stunt virus with four along the edges and ninety-two subunits; a newly described virus from a species of red spider (Arachnida) appears to have three and forty-two subunits respectively, the virus of turnip yellow mosaic has two subunits along the edge with one in the middle giving a total of thirty-two; this virus appears not to be a true icosahedron. Finally, there is the bacteriophage ϕX174 with two along the edge and a total of twelve subunits.

It is thus possible to form a sequence of these very small viruses by counting the protein subunits, and it emphasizes how similar these viruses may be in their make-up although coming from such totally dissimilar hosts. In the foregoing sequence the hosts consist of man himself, two species of plants, a member of the Arthropoda and a bacterium.

When we come to consider a large icosahedron such as the *Tipula* iridescent virus, which measures about 140 mμ, the number of subunits is of course correspondingly greater; in this case the number is computed to be 812.

We have seen how large a part the helix plays in the construction of the rod-shaped viruses, but apparently it is also concerned in the make-up of some of the larger spherical or near-spherical viruses. Preparations of three of these, the Newcastle disease virus (NDV), mumps and Sendai viruses have been examined on the electron microscope by Horne and Waterson using the negative-staining technique. These viruses are of variable size, 100–600 mμ, and spherical in shape; they have no regularly arranged subunits on the surface similar to those we have just discussed. When these viruses are partially disrupted, hollow strands, which form the internal component, are released. In broken portions of these strands, high magnification reveals a helical structure with a regular periodicity of 50 Å and a central hole with a diameter of 40 Å. Horne and Waterson point out the striking resemblance of this inner component to the particles of tobacco mosaic virus, and suggest that this RNA-bearing structure might be built from a number of subunits arranged in helical array, but forming a flexible structure similar to the filaments observed in preparations of the beet yellows virus.

The substructure of a bacterial virus, one of the T-even phages with a 'head' and 'tail', has been beautifully analysed by Brenner, Horne and their co-workers. They used a combination of methods for breaking up the phage-particles into their component parts, which were then examined by negative staining on the electron microscope. The phages which at one time were thought to be relatively simple particles of nucleoprotein are now shown to be almost fantastically complicated (plate x).

The general appearance of a T-even phage-particle can be seen in the diagram (fig. 1); there is the six-sided 'head' built of a large number of subunits, the contractile sheath with its core and the tail-fibres. Much structural detail of these component parts is shown in electron micrographs of material negatively stained. The technique shows the cores to be hollow and also shows the helical nature of the sheath; this has been calculated to consist of about 200 subunits. In end-on views of the contracted sheaths, about fifteen 'cog'-like subunits can be seen; if these represent the subunits in one turn of a helix the entire sheath should contain thirteen such turns. An outstanding feature of the tail-fibres is the characteristic kinking at the centre.

Internal structure

Improvements in the technique of cutting ultrathin sections not only allow sections to be cut of the virus particle itself but also of serial sections through one particle. This obviously is one way of examining the internal structure of the virus, but the difficulty here is that the electron microscope cannot differentiate between the chemical constituents of the virus. Sections do, however, reveal differences in structure and quantities of contents and also show up the empty particles.

Another method is to treat the virus particles with enzymes which dissolve away the protein covering and leave behind the nucleoprotein.

For studying the internal structure of virus particles by ultrathin sectioning it is, of course, better to choose the larger viruses, although this technique, using a diamond knife, has been applied to tobacco mosaic virus.

Ultrathin sections of vaccinia and fowl-pox viruses reveal a limiting membrane, a dense eccentrically placed body and a finely granular matrix. Sections of these viruses at a later stage seem to reveal developmental changes in which an elongated body surrounded by dense material can be seen.

Thin sections of the *Tipula* iridescent virus reveal a variety of different forms, ranging from apparently empty shells to the mature virus with its characteristic hexagonal shape.

The rickettsiae, which strictly speaking should not be regarded as true viruses, are very suitable for this type of study. Thin sections of purified *Rickettsia burnetii* reveal a limiting membrane within which lie a granular region and a dense central body. The configuration of the central body suggests that it may consist of an elongated and irregularly twisted strand.

The method of digesting the protein covering and observing the residue by means of the electron microscope has only been applied to the larger viruses. When the influenza virus was heated on the microscope grid with various enzymes, the nucleoprotein component was left behind in the form of an irregular ring built of long threads. A similar result has been obtained with the virus of Newcastle disease of fowls; when stained with phosphotungstic acid the ring showed a number of dark lines, possibly representing the tracks along the protein to which the long nucleic acid molecule is bound.

Behaviour of viruses in the cell

The technique of ultrathin sectioning has allowed the virus to be observed *in situ* in the cell, and a good deal of information on its behaviour has been obtained. Perhaps the least successful in this direction have been the studies of viruses in plant cells. This is partly due to the difficulty in distinguishing between the virus and the normal cell constituents and is particularly true of the very small plant viruses. It occasionally happens, however, that the virus may be concentrated in a small area and then becomes easily visible. In ultrathin sections of *Datura stramonium* infected with tomato bushy stunt virus, micro-crystals of the virus may sometimes be seen.

In cutting ultrathin sections of tobacco leaves affected with the mosaic virus, Nixon chanced to obtain a complete longitudinal section through a hair cell. By estimating the volume of the section and that of the whole cell from which it was cut the number of virus particles in the cell could be calculated. The micrograph showed that virus particles were the main component of the cell contents, and the calculation indicated that the cell contained about 6×10^7 particles.

It is much easier to observe the animal viruses in the cells, particularly when grown in tissue culture. Certain insect viruses, too, are very suitable for observation in thin sections of infected tissue.

It was shown by Morgan and his colleagues, by means of sections of chorioallantoic membrane cells infected with herpes virus, that the virus differs slightly according to whether it is in the cytoplasm or the nucleus of the cell. The intracellular cytoplasmic particles consist of a dense central body, surrounded by a double membrane.

In the nucleus, however, the particles contained the same central body but only one membrane. This does not, however, seem to be an invariable occurrence, as particles with double membranes are sometimes found in the nuclear matrix.

In the later study of a particular strain of herpes virus Morgan and his colleagues suggest that the virus differentiates near the nuclear margin and that the particles become enclosed within a second peripheral membrane while still within the nucleus. They have observed reduplication of the nuclear membrane itself, which suggests a method by which the virus particles can be extruded into the cytoplasm without rupturing the nucleus. Reduplications of the nuclear membrane are deposited behind the virus, and this allows the virus particles enclosed in sacks formed by the membranes to be extruded without rupture of the cell.

Another type of peculiar membrane formation has been observed in the fatbody tissue of the *Tipulid* larva infected with the iridescent virus. In this case, however, the nucleus is not involved since the virus occurs only in the cytoplasm; groups of virus particles occur, varying from one or two to fifty or more, enclosed in membranes which are reduplicated three or four times so that in sections the virus particles appear surrounded by three of four concentric rings. The significance of this peculiar phenomenon is at present obscure, though it may possibly be a defensive mechanism to ‘wall off’ the virus particles.

5

THE VIRUS IN THE CELL

Multiplication of Viruses. Tissue Culture of Viruses

Multiplication of viruses

In considering how viruses multiply or replicate themselves it is best to avoid any comparison with the growth of an organism. From a survey of what is known of replication in the viruses of plants and the higher animals a general picture of biosynthesis emerges, an assembly rather than a multiplication. Luria puts it like this—'virus multiplication belongs on the level of the replication of sub-cellular elements', or according to Pirie 'it is the exploitation and diversion of the pre-existing synthetic capacities of the host cell'.

In the replication of many viruses, especially the smaller ones, a dual process is involved in which the protein and nucleic acid are formed separately and then polymerized into the infectious particle. The new advances in the techniques of electron microscopy and X-ray diffraction studies (described in chapter 4) have given us a greater knowledge of the ultrastructure of virus particles. Using this, it is possible, to a certain extent, to visualize the process of its construction or replication.

It was shown by Markham and Smith that the purified virus of turnip yellow mosaic would separate, after centrifuging, into two portions called the 'top and bottom components'. On the electron microscope the two kinds of particles looked similar, but the top component was of lower density and contained no nucleic acid. Moreover, infectivity was confined to the bottom component which did contain the nucleic acid. This was the first definite proof that the infectivity of a plant virus was centred in the nucleic acid of the virus particle and that the nucleic acid was inside the protein coat. Since then a number of other viruses has been shown to have top and bottom components, and empty shells similar to the 'top component' have also been observed in several animal viruses.

Tobacco mosaic virus

A great step forward was made by Gierer and Schramm and Fraenkel-Conrat when they showed independently in 1956 that the RNA of tobacco mosaic virus (TMV) could initiate infection in the absence of the protein part of the virus particle. It has since been shown by Fraenkel-Conrat and Williams that the protein and the RNA can be separated and the two repolymerized, or the protein can be polymerized by itself. In the latter case some controlling factor seems to be lost as the resulting rods grow larger than the normal size of TMV particles ($350\ m\mu \times 15\ m\mu$). When the protein and the RNA are polymerized the infectivity of the resulting rods is about 60 per cent of the infectivity of normal TMV. The whole process has been carried out *in vitro* by Takahashi, whereby infectious TMV has been reconstituted with an anomalous non-infectious protein isolated from mosaic-diseased tobacco plants and with RNA isolated from TMV.

It has been observed on several occasions that after a virus enters a susceptible cell there ensues a period, called the 'eclipse phase', when the virus is undetectable. Furthermore, the eclipse phase is shorter when nucleic acid is inoculated by itself into the plant than when whole virus is used. This is what would be expected if during the eclipse phase the nucleic acid has to be denuded of its protein coat before replication can begin. It is possible that the protein merely protects the nucleic acid from the enzymes in the cell.

We can therefore visualize a kind of sub-assembly system for making the TMV; the subunits are made on the information contained in the RNA and then more and more subunits are made. It appears to be easier to replicate these small subunits and then to build them up than to build up the virus particle with larger units. For the smaller viruses with icosahedral symmetry Klug calculates that sixty identical subunits is the largest number which can be packed as a nearly spherical shell.

Bacterial viruses

The most detailed picture of virus replication, however, has been obtained from the studies of the T_2 and T-even bacterial virus or phages. To follow this it will first be necessary to describe in some detail the structure of the virus particle, because the structure is very much concerned in the beautiful mechanism of replication. The general shape of the phage-particle is like that of a tadpole with a 'head' and a 'tail' and in the electron microscope both are easily distinguished. The tail has been found to be composed of an inner core and an outer cover, and it can again be differentiated into a distal and a proximal part; the distal part seems to be composed of fibres which appear to be appendages to the tails. The tails end in a kind of tail-plate to which these fibres are attached (fig. 1). Anderson first showed how the phage-particle infects the cell. Contrary to what might be expected, the virus attaches itself tail first to the bacterial cell wall which it then penetrates, allowing the DNA contents of the head to pass down into the bacterium.

All the different component parts of a phage-particle have been isolated and beautifully illustrated by Brenner *et al.* and one of their electron micrographs is reproduced here (plate x).

In an infection of the bacterium *Escherichia coli* by T_2 phage, Kellenberger and his co-workers have shown that the breakdown of the bacterial nucleus is accompanied by the formation of marginal vacuoles containing DNA, from which the pool of phage DNA develops some minutes later. The phage DNA starts to increase 6 to 7 min after infection. Then, about 9 min after infection, the protein precursors (that is immature phages) appear and, finally at 13 min, complete phages appear in the infected cell.

Fowl-plague virus

This belongs to the myxovirus group and seems morphologically similar to influenza A virus; the multiplication cycle has been studied by Werner Schäfer from whose work this account is taken. The fowl-plague virus (FPV) is a sphere measuring 70–80 mμ in diameter; a model of the particle is illustrated in fig. 2. Unlike the small plant viruses, which consist only of RNA and protein, FPV is composed of protein, carbohydrates, lipids and RNA, the lipid content being about 25 per cent and the RNA about 1·8 per cent.

The cycle of replication is thought to be somewhat as follows: the infective virus particle containing what is called the *g*-antigen, which

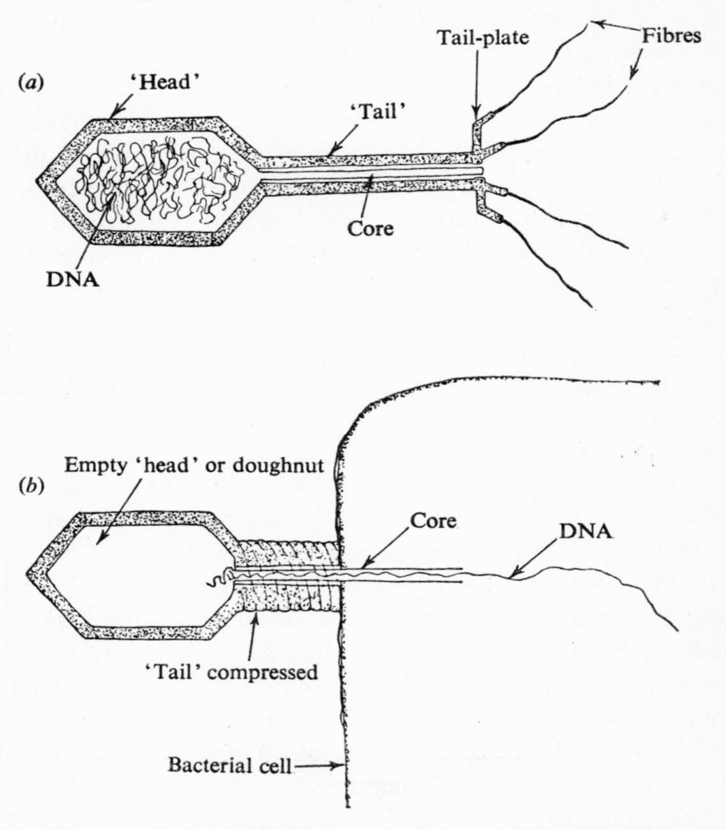

Fig. 1. (*a*) Diagrammatic representation of a T_2 bacteriophage particle showing the six-sided 'head' containing the strands of deoxyribonucleic acid, the 'tail' with its central core, the 'tail'-plate and fibres. (*b*) The same bacteriophage particle attached to the bacterial cell; the 'tail' has contracted, the central core has penetrated the wall of the bacterium, and the deoxyribonucleic acid has passed down into the cell.

is the viral nucleoprotein, haemagglutinin and lipids, becomes attached to the host cell. The particle then breaks down and releases the central mass, that is, the viral nucleoprotein, which enters the cell; the fate of the haemagglutinin and lipid at this stage is not yet known.

PLATE I

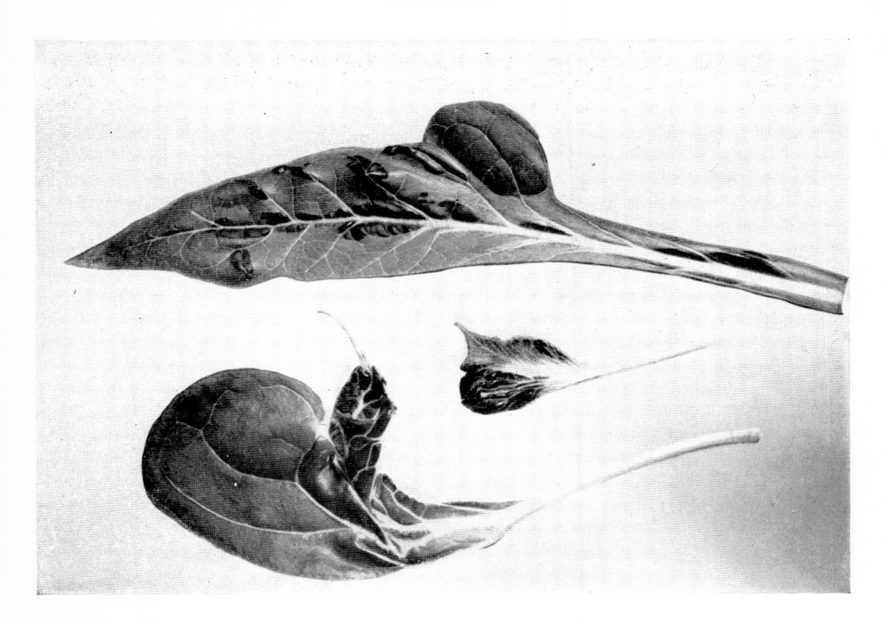

(A) Tobacco mosaic; the virus causing this disease was the first to be discovered.

(B) Tulip break, the oldest known plant-virus disease. (A.R.C. Virus Research Unit.)

PLATE II

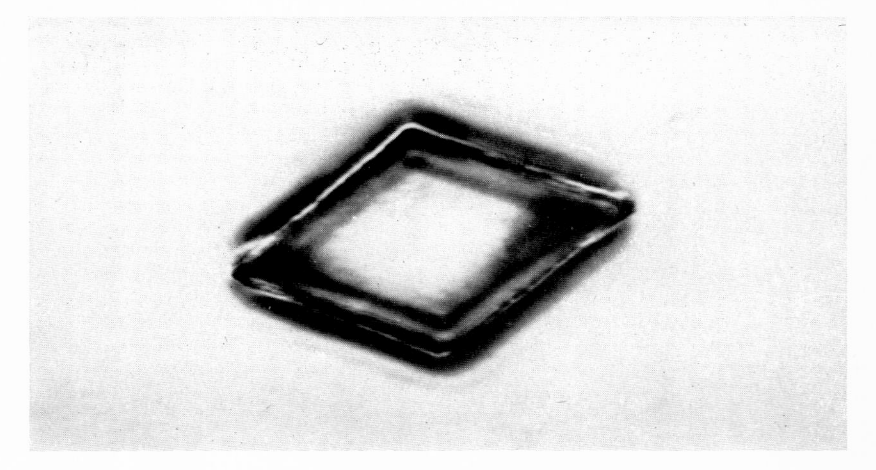

(A) Light micrograph of a crystal of turnip-crinkle virus. ×392. (A.R.C. Virus Research Unit.)

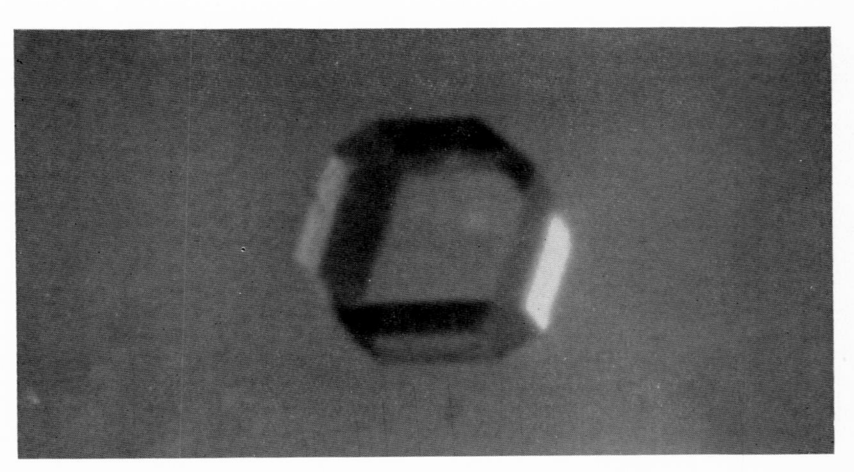

(B) Light micrograph of a crystal of poliovirus. ×425. (Virus Laboratory, University of California, Berkeley.)

PLATE III

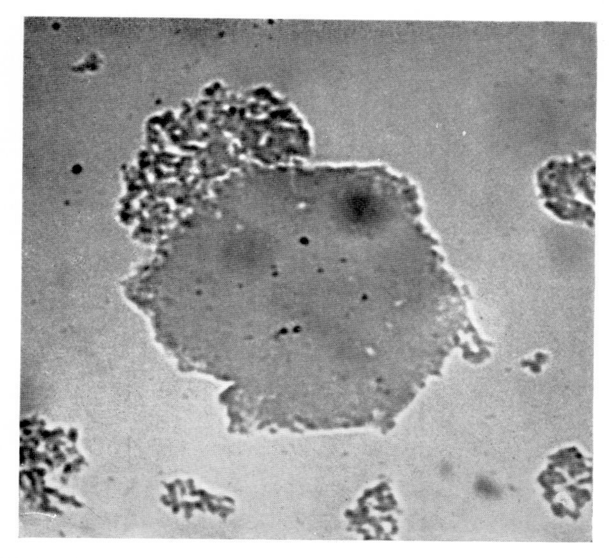

(A) Light micrograph of a crystal of the *Tipula* iridescent virus. ×2150.

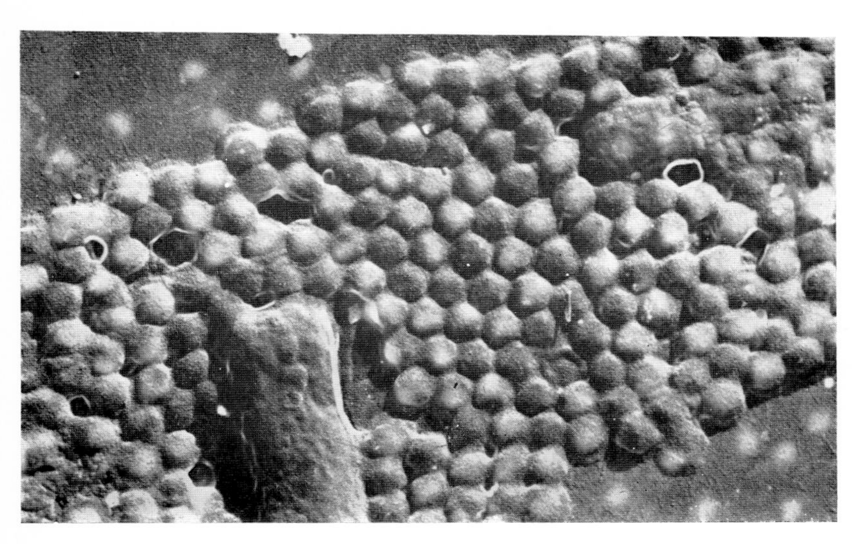

(B) Electron micrograph of a carbon replica of a microcrystal of the *Tipula* iridescent virus. ×42,500. (A.R.C. Virus Research Unit.)

PLATE IV

Electron micrograph of a crystal of the rhombic type of tobacco necrosis virus, together with an orthorhombic crystal model for comparison. ×144,200. (Courtesy Louis W. Labaw.)

PLATE V

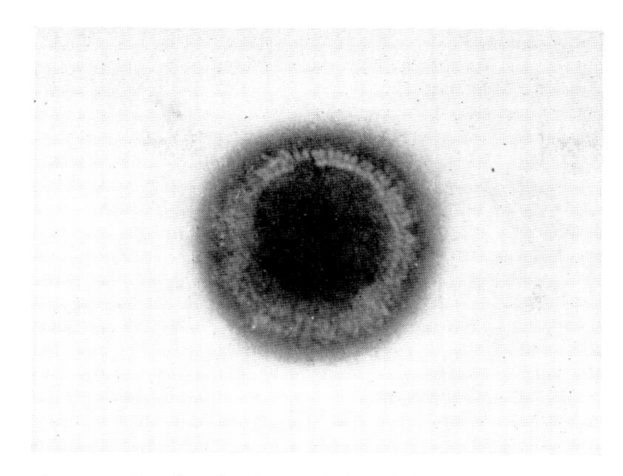

(A) Electron micrograph of a single particle of the *Tipula* iridescent virus; note the subunits forming the outer protein coat. × 198,750.

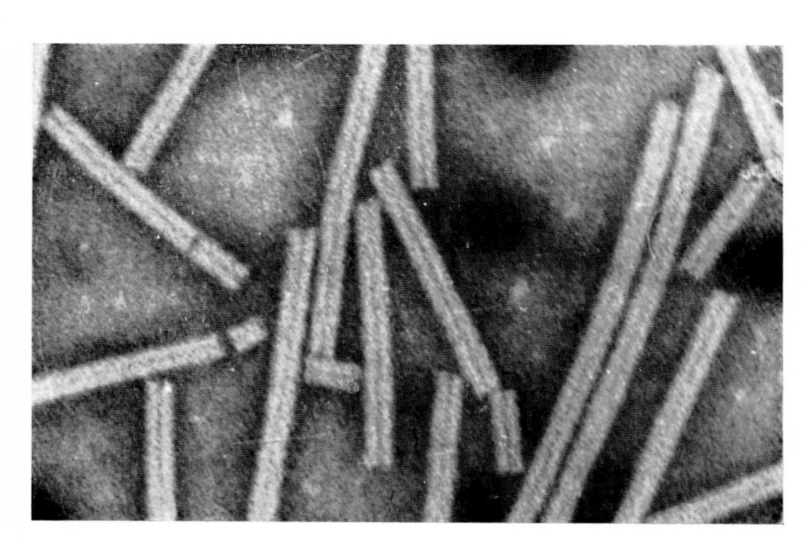

(B) Electron micrograph of tobacco mosaic virus particles stained with phosphotungstic acid; note the hollow centre picked out by the stain. × 265,000. (A.R.C. Virus Research Unit.)

PLATE VI

(A) A model of an icosahedron, shadowed by two light sources and orientated so that an apex of the hexagonal contour points directly towards each light source.

(B) A frozen-dried doubly shadowed particle of the *Tipula* iridescent virus, orientated similarly to the icosahedral model shown in (A) and exhibiting similar shadows. $\times$ 54,000. (After Williams and Smith.)

PLATE VII

Model of a tobacco mosaic virus particle, showing the protein subunits set in a helix; some of the subunits have been removed to show the strand of ribonucleic acid. (After Franklin and Klug.)

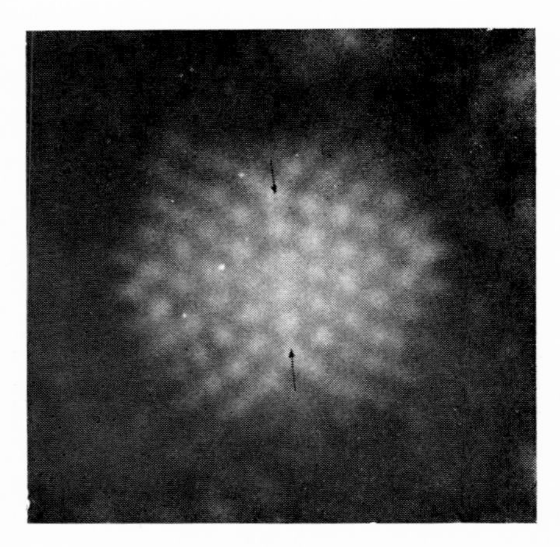

(A) Adenovirus particle at high magnification (×450,000); note that the subunits have a definite pattern.

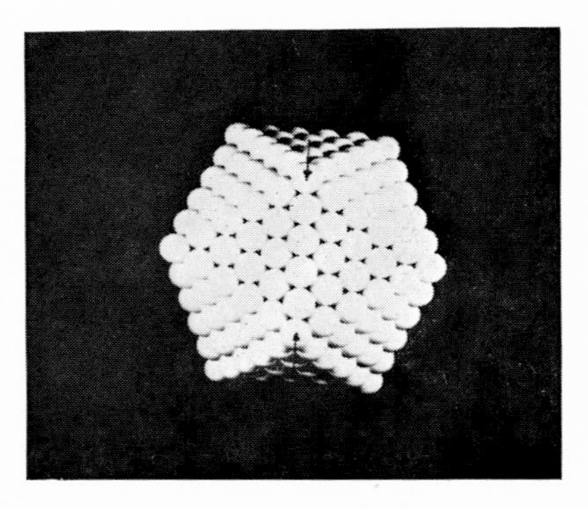

(B) Photograph of a model of an icosahedron the in same orientation. The arrows mark subunits which lie on fivefold rotational symmetry axes. (After Horne, Brenner, Waterson and Wildy.)

PLATE IX

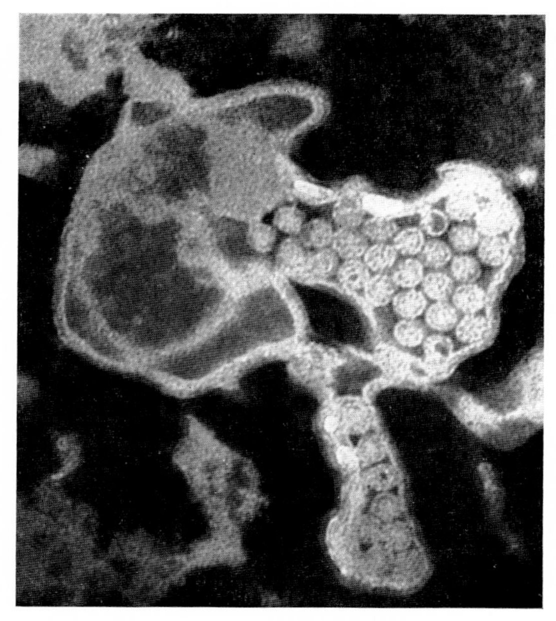

(A) Body showing poliovirus particles in later stage of assembly at $5\frac{1}{2}$ hr. Subunits are clearly visible in both complete and incomplete particles. ×127,500. (R. W. Horne and J. Nagington, *J. Mol. Biol.* **1**, 1959.)

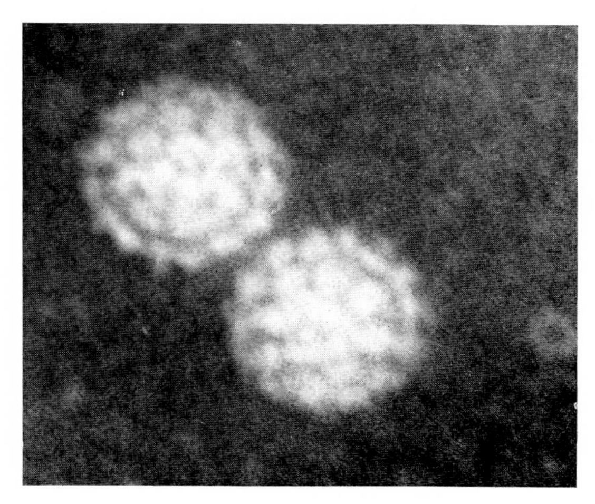

(B) Two poliovirus particles at a high magnification showing the arrangement of the subunits. ×825,000. (R. W. Horne and J. Nagington, *J. Mol. Biol.* **1**, 1959.)

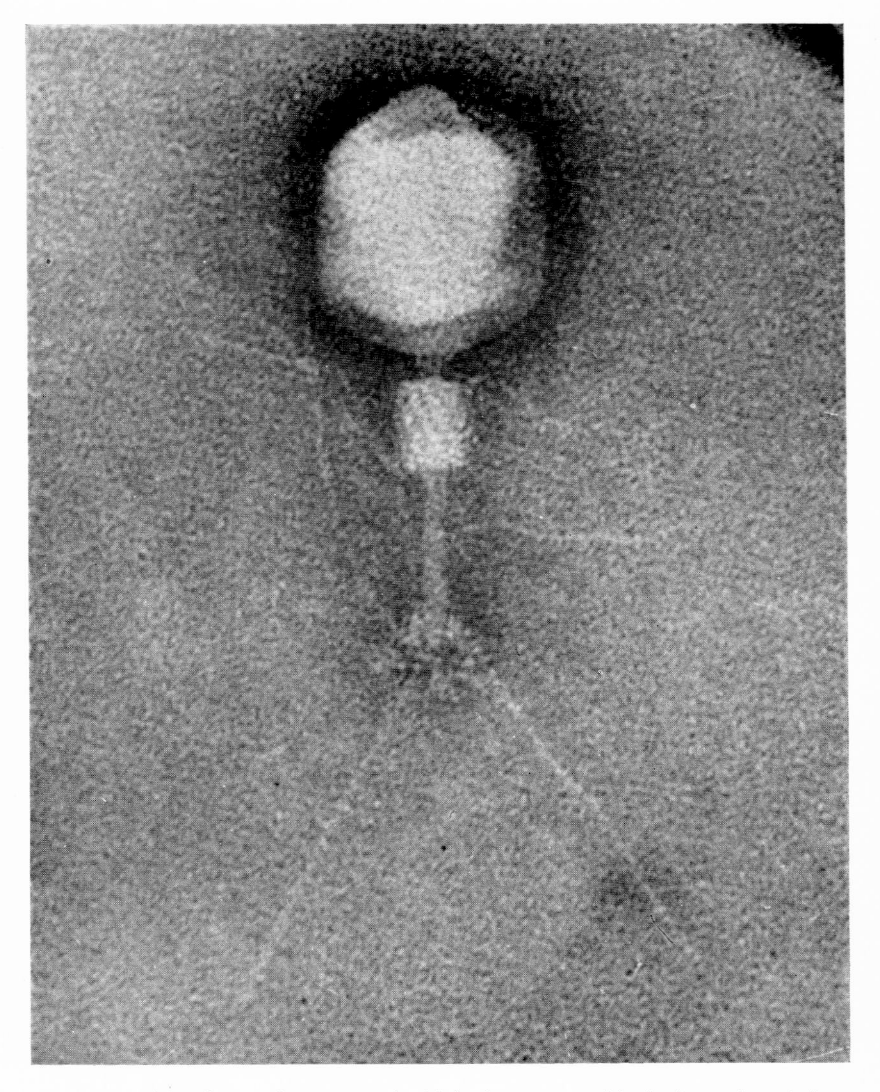

Isolated particle of a T_2 phage, treated with hydrogen peroxide, it shows the relations of the filled head, contracted sheath, core and tail-fibres. (PTA embedded. $\times$ 391,500.) (After S. Brenner, G. Streisinger and R. W. Horne *et al.* *J. Mol. Biol.* **1**, 1959.)

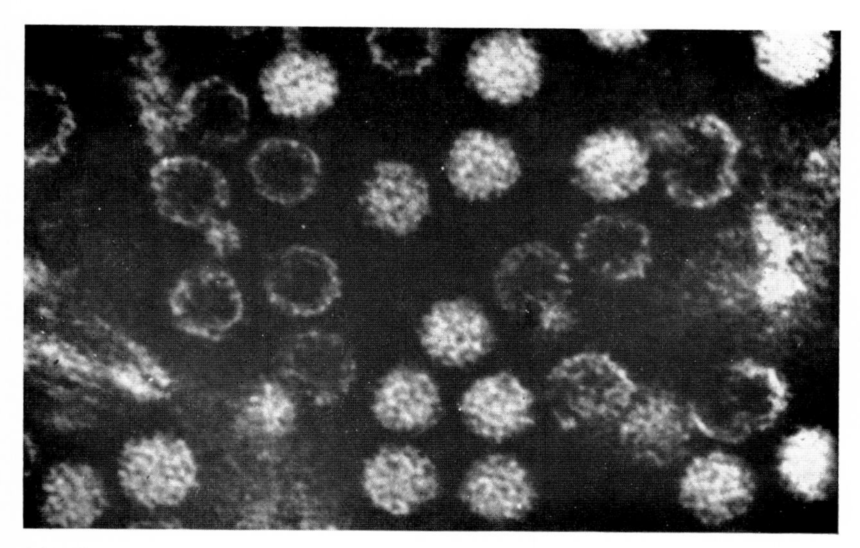

(A) Electron micrograph of turnip-crinkle virus particles, negatively stained with phosphotungstic acid; note the protein subunits and some empty shells. ×296,000.

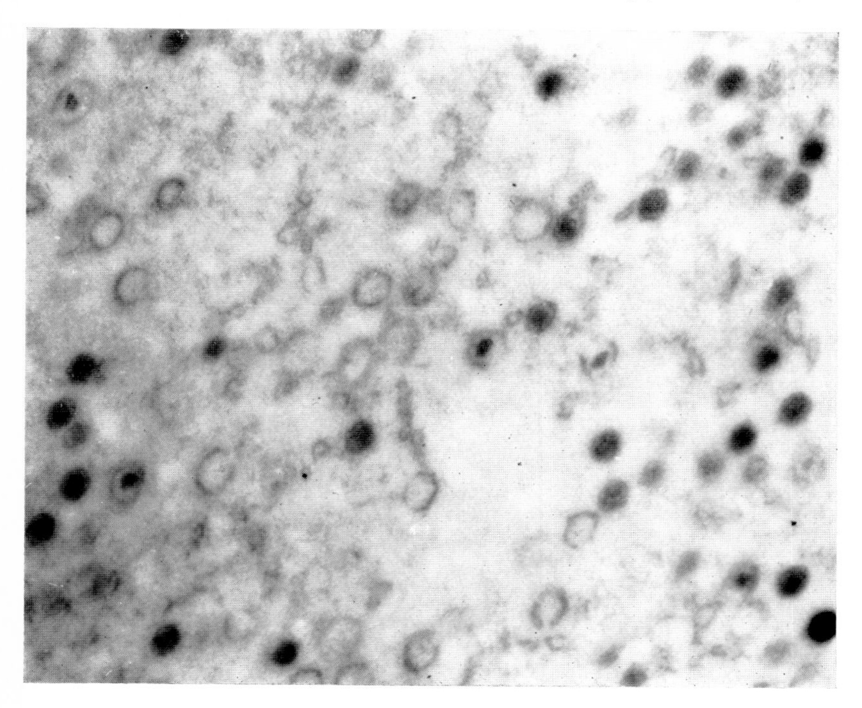

(B) Electron micrograph of part of a section through the fatbody of a crane-fly larva at an early stage of infection with the *Tipula* iridescent virus; note the empty shells, half-filled and mature virus particles. ×36,750. (A.R.C. Virus Research Unit.)

PLATE XII

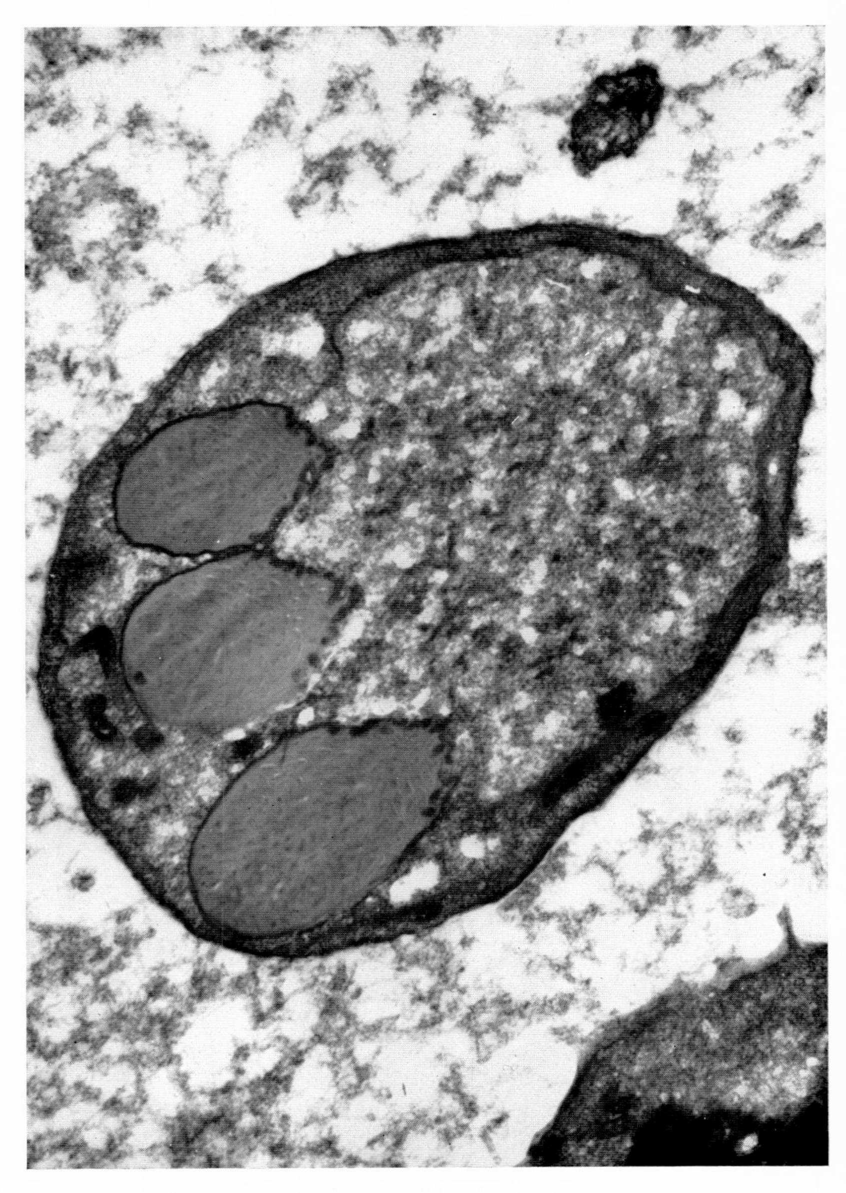

Electron micrograph of a section through a blood-cell of a crane-fly larva infected with a nuclear polyhedral disease; note the three polyhedra with the rod-shaped virus particles arranged in a definite order as they become occluded in the crystal. At the apices of the crystals can be seen those virus particles too late to be enclosed; they are stained black by the osmic acid. Some loose virus-rods are still visible in the nucleus of the cells. ×10,000. (A.R.C. Virus Research Unit.)

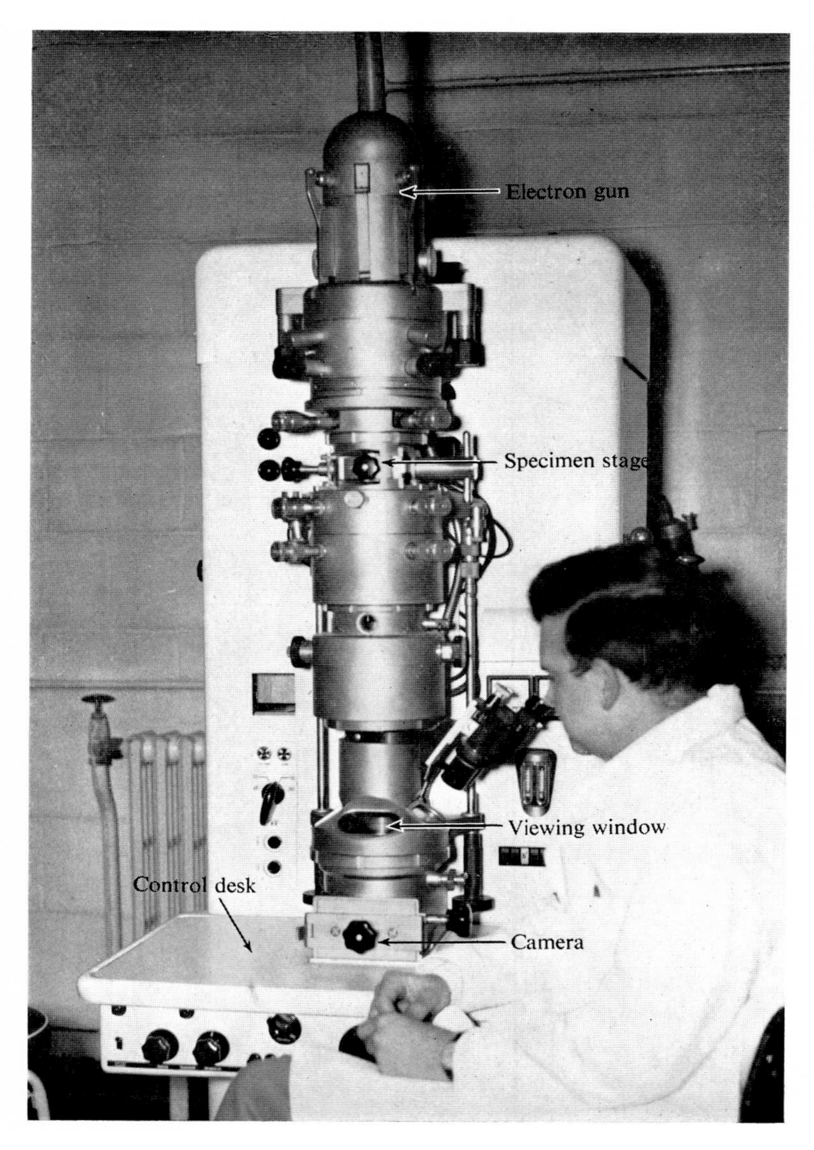

Electron microscope; Siemens Elmiskop I, showing the general appearance of the instrument. (A.R.C. Virus Research Unit.)

PLATE XIV

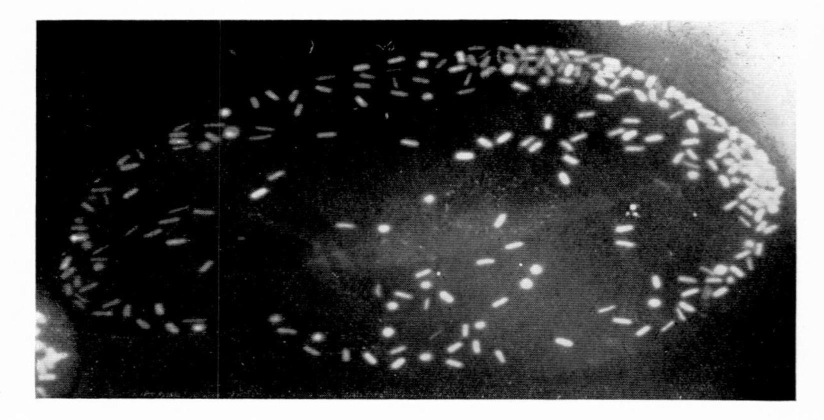

(A) Electron micrograph of a *nuclear* polyhedral crystal from a larva of *Lymantria monacha*, the nun moth. The crystal has been dissolved in weak alkali, leaving behind the rod-shaped virus particles in the enclosing membrane. × 10,000.

(B) Electron micrograph of a *cytoplasmic* polyhedral crystal from a larva of *Phlogophora meticulosa*, the large angleshades moth, treated similarly to that in (A). Note the round holes in the residue of the crystal which contained the spherical virus particles and the absence of a membrane. × 25,000. (A.R.C. Virus Research Unit.)

PLATE XV

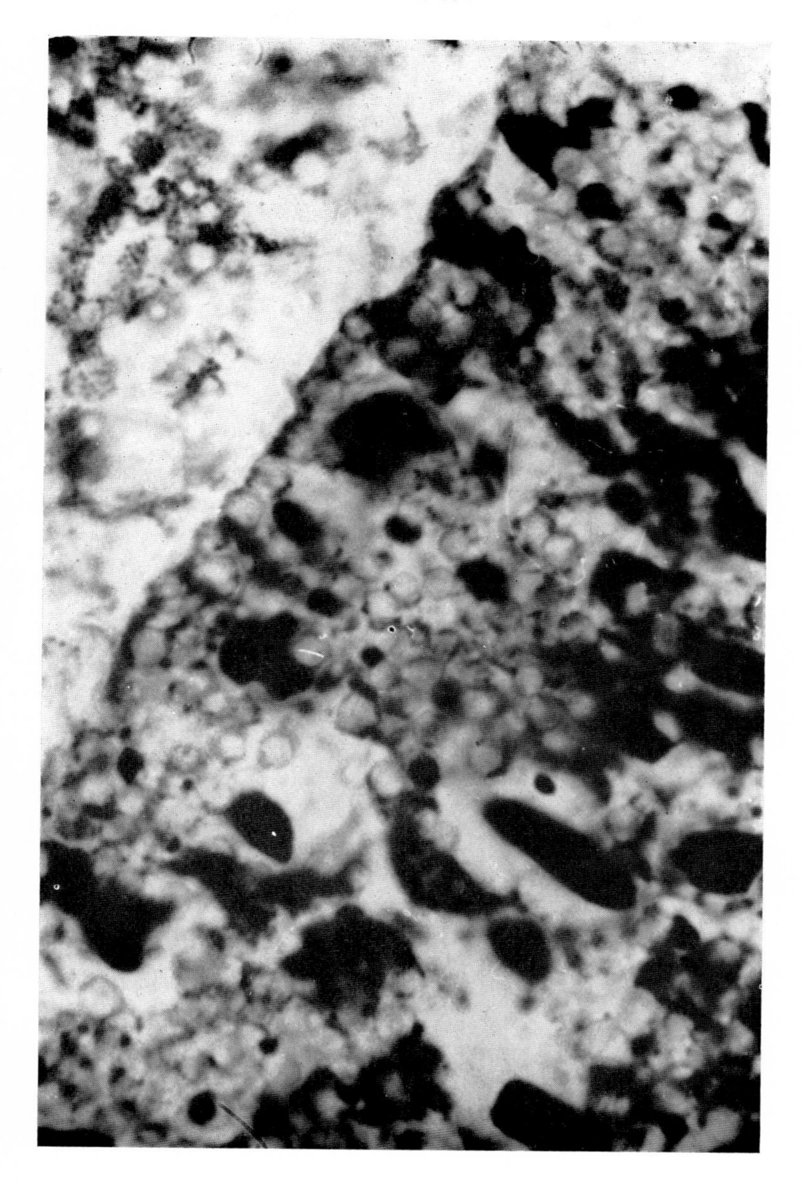

xv. Electron micrograph of a section through part of the mid-intestine of an apparently healthy fritillary butterfly, *Argynnis dia* (L.) showing large numbers of polyhedral crystals. $\times 1200$. (A.R.C. Virus Research Unit.)

PLATE XVI

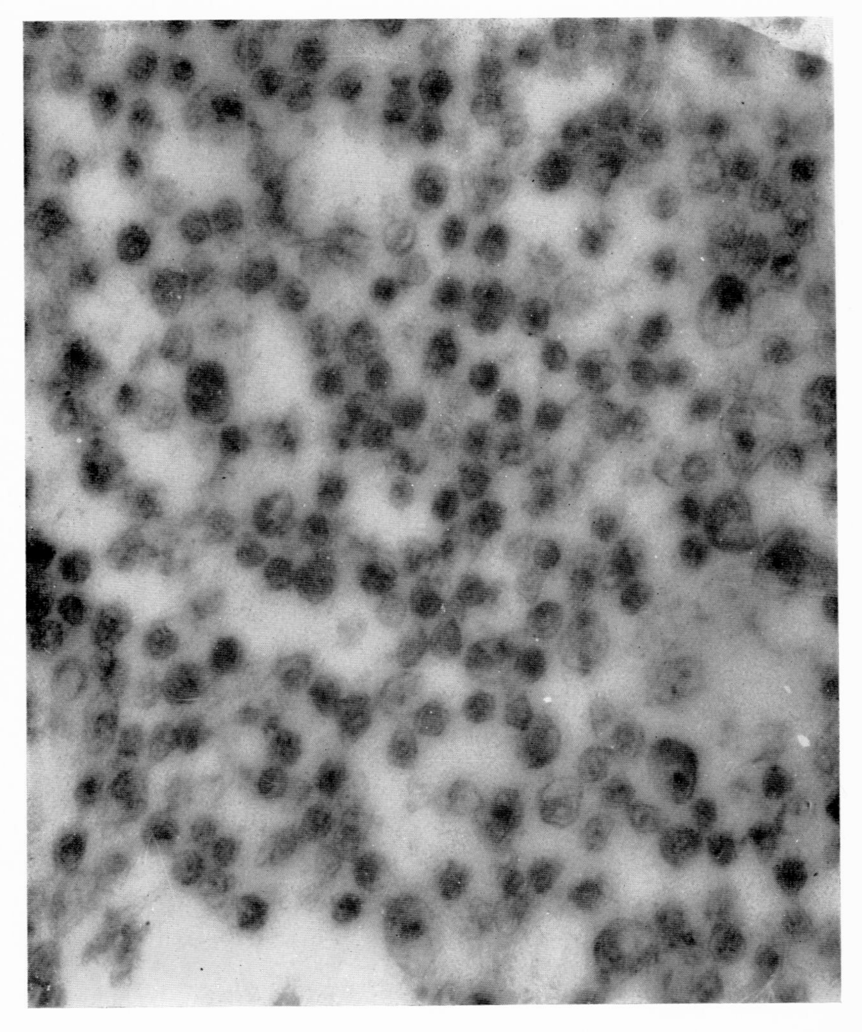

Electron micrograph of virus particles in spontaneous mouse leukaemia. $\times$ 30,000. (Courtesy Leon Dmochowski.)

The material liberated into the cell seems to induce the cell nucleus to form new virus material, as after about 3 hr there is an increase of a substance which behaves biologically like the viral nucleoprotein (*g*-antigen). Shortly after the appearance of the new *g*-antigen in the nucleus there is an increase in the haemagglutinin which is localized in the cytoplasm. These materials seem therefore to be produced in separate systems of the cell, the viral nucleoprotein in the nucleus and the haemagglutinin in the cytoplasm. It is not clear what induces the formation of these systems but, by analogy with tobacco mosaic virus, we can suppose that the RNA can initiate the production of the viral nucleoprotein.

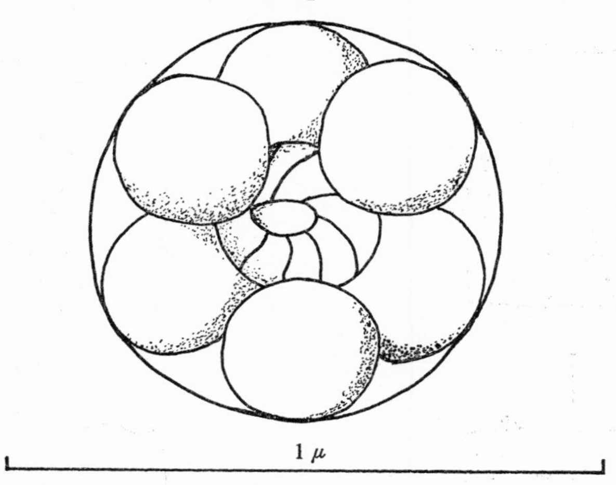

Fig. 2. Model of infectious particle of fowl-plague virus. (After Werner Schäfer, *Perspectives in Virology*.)

The two virus components seem to diffuse from their sites of formation to the cell margin, where they are finally assembled into the new infective particles. It is thought that in these units the components are held together by lipids from the cell wall. Filaments and short rods extending from the cell surface have been observed in the electron microscope (see fig. 3).

Herpes simplex virus

Morgan and his colleagues in the U.S.A. have made a careful study of the development of herpes virus in tissue cultures of HeLa cells by means of thin sections examined with the electron microscope. According to their interpretation, the first stage in the multiplication of the virus, which takes place in the nucleus, is the appearance of large numbers of uniformly shaped granules measuring about 20 mμ in diameter. These granular areas are thought to be foci in which the various components of the viral particle become spatially arranged.

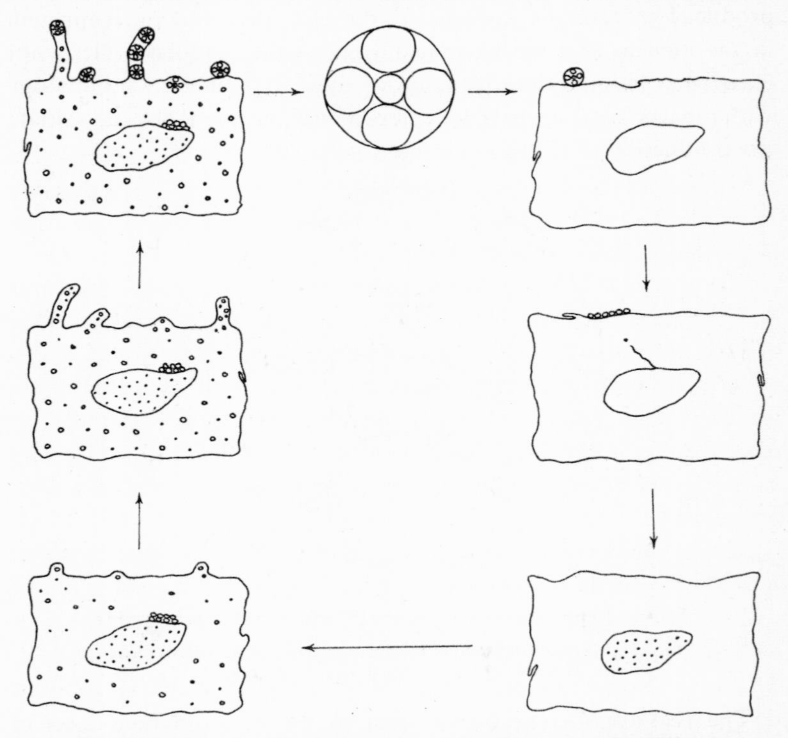

Fig. 3. Multiplication cycle of fowl-plague virus. (After Werner Schäfer.) The infectious particle attaches itself to the cell-wall, it then breaks up into subunits which enter the cell and induce the nucleus to form new virus material. Incomplete virus particles then pass into the cytoplasm and come together to form the mature virus particle which is then extruded from the cell in filaments to start the cycle over again.

This involves the progressive development of a membrane which encloses the central body of the virus particle. In one section of a nucleus this can apparently be seen in progress; many membranes seem to be actually in process of formation at many points in the aggregate of granules. When the herpes virus particle is first formed it has only a single membrane, but later a second membrane develops, and after the formation of this there appears to be a change in the central body which becomes uniformly opaque.

This mode of replication seems somewhat similar to what has been observed in the multiplication of the *Tipula* iridescent virus; in this case, however, the membranes seem to form first and areas of isolated granules have not been observed. Later, groups of six granules or spheres are seen in the centre of the membrane and these appear to coalesce to form the central body.

Insect viruses

Sanders has pointed out that an abundant field exists for the investigation of the association between animal viruses in their intracellular development and the well-defined cell organelles. There is the cell nucleus which contains both DNA and RNA, the latter being mainly concentrated in the nucleus. In the cytoplasm there are the mitochondria, which are rod-shaped or rounded, and which contain phospholipid and protein with possibly a little RNA. In addition, there is the endoplasmic reticulum which is rich in RNA and may correspond in the unbroken cell to the microsomes which can be centrifuged out of cellular homogenates and have probably been mistaken more than once for viruses.

Many cellular organelles are themselves capable of replication and it is interesting to speculate whether viruses are able to make use of existing cell structures during the process of their own multiplication. In the case of a virus known as the *Anopheles* A virus, an association between it and the endoplasmic reticulum of the cell has been suggested by Friedlander.

An interesting phenomenon, the exact significance of which is not fully understood, has been observed with a number of viruses from both plants and animals. This is the occurrence of fully formed but empty virus membranes or shells; they have been observed in the viruses of turnip yellow mosaic, of influenza, of poliomyelitis, of a frog kidney tumour and in the insect virus known as the *Tipula* iridescent virus (TIV). They appear to differ from the empty heads of bacterial viruses (fig. 1 (*b*)), since these are known to be left outside the bacterial cell after the injection of the DNA.

Smith has suggested that these empty shells of TIV may represent a stage in the multiplication of the virus. It is in the development of this virus that a very close resemblance has been observed between the normal endoplasmic reticulum of the cell and the empty membranes of the developing virus. Indeed it is often difficult to distinguish between them on the electron microscope. These empty membranes or shells only occur in large numbers in the early stages of the disease, but they show considerable variation in their contents. Some may be completely empty, others contain a few strands of material which may be DNA, whilst others again may contain a doubly staining central body, and so on up to the presumably mature virus particle. Plate XI, B, is an electron micrograph of an ultrathin section of the fatbody from a larva of *Tipula paludosa* infected with TIV. All graduations can be seen from completely empty shells to the mature virus particle. On the evidence so far available, therefore, it looks as if the protein shell and the virus DNA are synthesized separately in the cell cytoplasm, but the method of entry of the DNA into the shell is not at present known.

In the fowl-plague virus we had an instance of an RNA virus in which the RNA is synthesized in the nucleus, but in TIV we have an example of a DNA virus in which the DNA is synthesized in the cytoplasm.

We saw in chapter 4 that TIV is an icosahedron. However, many insect viruses are rod-shaped and the situation as regards their mode of replication is a matter of controversy.

Some workers favour the idea of a complicated life-cycle consisting of numerous developmental forms, much on the lines of an organismal multiplication. Bergold arranged what he thought was a probable sequence of numerous particles found in preparations of insect polyhedral diseases under the electron microscope, and he proposed the following sequence of events for the rod-shaped viruses of the polyhedroses and granuloses of insects (see chapter 8). Their development begins with one or several small spheres (about 20 mμ in diameter) that grow within the 'developmental membrane'. These spheres elongate to kidney- and then to V-shaped forms, and appear finally as straight rods still within the developmental membrane. Sometime during this process each rod becomes surrounded by the intimate membrane. The rods with both membranes constitute the infectious virus that attaches to the host cell and nuclear membrane by a thin protrusion. In an unknown way, spherical subunits are released from the intimate membrane; these spheres begin the cycle again.

Similarly Bird suggests the following cycle of development for polyhedrosis viruses; polyhedra are ingested by a larva and dissolve in the gut, liberating virus rods. The rods attach to mid-gut cells and release infectious subunits; these pass through the mid-gut epithelium which in Lepidoptera is not susceptible to infection, and enter a nucleus of a susceptible cell. Multiplication takes place within the nucleus, and spherical bodies are formed; each sphere contains 'germs' for a number of virus particles, the mature form of which is rod-shaped. Bundles of rods, still contained by their developmental membranes, are surrounded by protein which crystallizes into polyhedral inclusion bodies. Rods from some of the bundles escape from their developmental membranes, are never occluded by polyhedral protein, and attach in large numbers to the nuclear membrane and also to other materials in the nucleus. They release subunits which penetrate and pass out through the nuclear membrane, and infect adjacent cells.

Such developmental cycles are unknown amongst the large number of viruses of the higher animals which have been studied and, if confirmed, would set the rod-shaped viruses affecting insects sharply apart from all other viruses, with the possible exception of the large agent of psittacosis.

In Cambridge we have approached the question of the multiplication of insect viruses from a slightly different viewpoint. We have endeavoured to deduct the method of replication from high resolution electron microscopy of the structure of the virus particles.

So far as the rod-shaped viruses are concerned we have not yet been able actually to observe the assembly of the new particles, although some of our observations on the ultrastructure of the rods may bear on this. In the breakdown of the virus rods from the nuclear polyhedroses the first stage in the liberation of the contents of the intimate membrane is the peeling off of the outer capsule. The capsules break in the centre and fold backward, thus forming two spheres still joined in the middle. These finally break apart and we consider them to be the same as Bergold's spherical 'subunits', which he said were discharged from the intimate membrane.

The intimate membrane is then exposed and it measures about 20 Å in thickness; it has a slightly different structure at either end. At very high magnification the contents of the intimate membrane give the appearance of a wide-spread helix; these contents are discharged from either end of the intimate membrane and appear to uncoil as they flow out. We consider that this helix is in part nucleic acid (DNA) and it differs markedly from the 'subunits' described by Bergold. The presence of a helix puts the insect rod-shaped viruses more in line with the ultrastructure of other viruses as described in some detail in chapter 4.

Tissue culture of viruses

One of the fundamental characteristics of viruses is their inability to grow in a cell-free or synthetic medium; they can only multiply in a living, susceptible cell. However, since the cell, and not the whole organism, is the nurse to a virus it follows that viruses can be grown in cells which are themselves being cultivated in a nutritive medium. This is what is meant by the tissue culture of viruses; development of this technique has been among the most important advances in virology of recent years and from it has directly arisen the hope of controlling such diseases as measles, poliomyelitis and the common cold.

Although tissue-culture methods have been known for over forty years, it was not until 1925 that a virus, vaccinia, was shown to have increased by about 50,000 times when cultured in rabbit testicular tissue.

A number of discoveries have helped to give the tissue culture of viruses its present importance, the main impetus being supplied by the urgent necessity of finding a vaccine for poliovirus. One important discovery was the recognition that viruses growing in cells produce changes which can easily be recognized. This allows for a rapid diagnosis of some viruses, especially poliovirus, and such diagnostic facilities are now available in the Public Health Laboratory Service of Great Britain. The recognizable 'cytopathogenic' changes in cell cultures of poliovirus have thus done away with the necessity for using large numbers of monkeys for tests and even as a source of virus.

Partly arising out of the discovery of these cytopathogenic changes is the recognition of large numbers of new viruses, the existence of which had never been suspected. The so-called adenoviruses, enteroviruses and new respiratory viruses have been discovered in this way, and a recent review has reported no less than seventy new viruses.

The discovery and development of antibiotics have proved of inestimable value to tissue-culture methods, since by their addition to the medium bacterial contamination is eliminated without effect on the virus.

Another great step forward was the development of very uniform cultures, consisting of a single sheet or monolayer of cells. This was made possible by the treatment of cell suspensions by trypsin. Recently Dulbecco and his colleagues in the U.S.A. have developed this method still further by the 'plaque technique'. They found that the spread of virus through the medium could be prevented by overlaying the cell monolayer with nutrient agar immediately after inoculation of the virus into the cells. Under these conditions only direct spread of the virus is possible, and this results in the development of plaques which become visible to the naked eye. Since each plaque represents one infective unit of virus, an accurate method of virus assay is obtained. Further refinements in this technique have recently been made by Porterfield and Allison using a monolayer of chick-embryo cells; by adding tris (hydroxymethyl) aminomethane to the agar overlay they have obtained well-defined plaques with vaccinia, ectromelia and herpes simplex viruses.

There are two main types of tissue culture; in one, fragments of tissue are used and in the other, outlined above, suspensions of dispersed cells are grown, mainly on glass.

The tissues used in the first type are obtained from freshly killed animals or from embryos, etc.; whilst stock cultures of cells are maintained in many laboratories from which subcultures can be obtained at any time. The nutritive medium in which the tissue cultures are grown consists essentially of blood plasma or serum or a mixture of the two, to which a balanced salt solution is usually added.

As might be expected, a cell culture consists of a heterogeneous collection of cells which do not react in a uniform manner to a given virus. Better results are therefore obtained when the cell culture is a 'clone' in which all the cells are derived from a single cell.

A great step forward in the culture of viruses pathogenic to man was made by Gey, who initiated a strain of cells known as HeLa cells. These came originally from a human cancer, cervical carcinoma tissue, and have been adopted for use in many laboratories.

Some interesting phenomena have been observed during this large-scale development in tissue culture. Thus sometimes a line of cells which is resistant to infection with a particular virus becomes, after prolonged cultivation *in vitro*, susceptible to the virus. For example, rabbit embryonic kidney cells which for long could not be infected with poliovirus eventually became susceptible.

We shall discuss in chapter 9 the phenomenon of latency or the 'carrying' of viruses without symptoms by infected animals or plants. In the cell culture of viruses a somewhat similar phenomenon seems to occur. Puck and Marcus exposed HeLa cells to large quantities of Newcastle disease virus, another name for fowl-pest, and found that it destroyed the majority of the cells. However, some cells survived apparently undamaged and clones were propagated from these. After two years these cells were shown to be still carrying the virus and to be able to destroy other lines of HeLa cells brought into contact with them.

Occasionally during the cultivation of cells an unsuspected virus may turn up in the culture, having been brought in by some cells already infected. This has sometimes happened during attempts to cultivate insect cells when the polyhedral crystals characteristic of some insect viruses have appeared in the cells.

The spontaneous development of malignant cells has also been observed during tissue-culture work, and the possible connexion between viruses and tumours is discussed in chapter 10.

In considering some of the practical applications of the tissue culture of viruses we have already mentioned some of the diagnostic uses. A quotation from Westwood may be appropriate here: 'When Enders and his colleagues demonstrated that poliovirus would multiply in non-neural tissues in culture, they provided at one stroke not only the main driving force behind much of the tremendous volume of tissue-culture research of the last decade, but also a simple, cheap laboratory host for poliomyelitis research. Where, previously, a monkey would provide a single observation, tissue cultures derived from it will now provide several thousand.'

One very important practical benefit that arises is of course the preparation of vaccines, made possible by knowledge of changes brought about in viruses during their cultivation in tissues. The important change is attenuation or loss of virulence by the virus for man; such attenuated viruses, although no longer capable of inducing severe disease, have not lost their property of invoking the production of antibodies and thus conferring immunity on the inoculated subject. The use of virus strains of attenuated virulence as 'live' vaccines has long been practised, the classical case being vaccination against smallpox. A modern example is the use of an attenuated strain for vaccination against yellow fever. This was first accomplished in 1937 by Theiler and Smith, who transferred a virulent strain of yellow fever virus for about 114 passages in chick-embryo tissue culture. They obtained the strain known as 17D, which has been used as a highly effective 'live' vaccine for millions of inoculations.

Similar attempts are being made to produce a 'live' poliovirus vaccine by the use of tissue-culture technique, but at present the danger of reversion to virulence is not sufficiently remote for its general use. This is not the same as the Salk vaccine at present widely used against poliomyelitis; the method used here is to inactivate the virus with formalin and, provided the inactivation is complete, this precludes any risk of reversion to virulence.

For a more complete account of the tissue culture of viruses the reader is referred to the accounts by Westwood (1959) and Enders (1959), while for a description of tissue-culture methods in general there is a little book by Willmer (1959).

6

HOW VIRUSES ARE SPREAD FROM HOST TO HOST

Higher Animals. Insects. Plants. Virus reservoirs and Sources of Infection

The methods of dissemination of viruses are as varied as their hosts and, whilst some are spread on fomites (any article in close contact with infection), many require an intermediary to act as vector. These vectors may be dogs or bats in the case of rabies, mosquitoes for yellow fever or equine encephalomyelitis, and many different types of plant-feeding insects for the plant viruses. The relationship between viruses and their vectors is a complicated and interesting one, and it is dealt with in some detail in the ensuing chapter. For the moment we are only concerned with the other methods of virus dissemination.

Higher animals

Probably the most important method of spread of animal viruses is the airborne route; it is the main means of dissemination of all respiratory infections and many others including smallpox and measles.

Airborne infection includes droplets expelled from the mouth or nose of an infected person or animal. When these droplets are small, 0·1 mm in diameter or less, they become so small by evaporation that before they reach the floor they can float in the air for many hours or even days as 'droplet nuclei'. These droplet nuclei may play an important part in the spread of some viruses. The virus of influenza, for example, on reaching the floor, bedclothes, and other objects, may survive on dust particles, and these may be subsequently resuspended in the air and inhaled by other persons. It has been demonstrated that influenza virus of the PR 8 strain could be recovered from dry dust exposed near a ferret infected with influenza; there is also a case on record of a research worker himself contracting influenza from a sneezing infected ferret. Between 1 and 10 per cent of the virus will withstand drying in household dust; under these conditions there is little depreciation after 3 days, 10 per cent may persist for a week, and 1 per cent for a fortnight.

There are examples of other viruses which may be airborne; there is the possibility that the virus of the common cold can be shaken from a handkerchief and travel a short distance through the air. Dried epidermal scales from cases of smallpox are regarded as infective and their aerial transfer may carry the infection to a distance. Experiments have shown that the viruses of poliomyelitis, influenza and laryngo-tracheitis can all be transmitted by the exposure of susceptible animals to artificially contaminated air. It has also been shown that the virus of Newcastle disease or fowl-pest can be recovered from the air of poultry houses containing infected fowls; air from an infected house sampled in 540- or 1080-litre quantities contained virus in sufficient concentrations to infect chick embryos.

Airborne fragments of feathers or excretions from parrots affected with psittacosis, if inhaled, are sufficient to infect the human subject.

One of the most infectious viruses is that of foot-and-mouth disease, which is spread by direct and indirect contact. The saliva, urine and milk of infected cattle are all infectious before the appearance of any symptoms, and it is probable that much spread of the disease takes place by indirect contact. When dried on certain materials the virus can remain infective for several weeks, and it can of course be carried around on the boots and clothing of farmworkers.

In fowl-pox, direct contagion is most frequent, but a small wound is apparently necessary for infection, such as abrasions of the mouth, which are common as the result of eating grit. Burnet found that it is necessary for grit to be present also in contaminated drinking-water before pigeons could be infected with the fowl-pox virus.

Hereditary transmission of viruses is not common but it does occur. Examples are the placental transmission to the embryo in such diseases as variola, varicella, Rift Valley fever and rinderpest. One of the most important cases of the hereditary transmission of a virus is that of breast cancer in mice. Some strains of mice may have an incidence of breast cancer in breeding females of 80 to 90 per cent, whilst in others the incidence is almost nil. In hybrids between two such strains it has been found that the incidence of breast cancer in the offsprings depends not upon the ordinary laws of inheritance but wholly upon whether the mother came from a high- or low-cancer stock. In other words the cancer of the breast in specially inbred strains of mice is caused by a virus found in the female mouse's milk.

Insects

The virus diseases of insects are dealt with in detail in chapter 8 and here we are only concerned with the methods of spread of the viruses. There seem to be two main infection routes, through the mouth and by hereditary transmission.

There is no doubt that the commonest mode of infection is by the ingestion of contaminated food, and this is mainly the food-plant, but there is also a tendency on the part of some species of caterpillars to eat the dead or moribund bodies of other caterpillars which have died of the virus infection. Contamination of the food-plant may be direct, by the excretions of infected larvae or their disintegrating bodies, or indirect by the carriage of the virus over a distance by the wind, birds or other insects. It has to be remembered that in two of the largest groups of insect viruses, the polyhedroses and granuloses, dispersal of infection is greatly aided by the fact that the virus particles are occluded in protein crystals and granules. These 'virus-containers' protect the infective agent itself from adverse conditions and allow it to remain infective for long periods, whilst at the same time facilitating its spread from plant to plant. Once ingested, the polyhedral crystals are dissolved in the gut of the insect and the virus particles are liberated.

The other important mode of spread of the insect viruses is by hereditary transmission; there is little doubt that in many cases the virus is passed through the eggs of the infected parent. There has been some controversy as to whether the virus is actually in the embryo or merely mechanically carried outside on the egg shell, whence it infects the young caterpillar on emerging. From the evidence, however, it now seems likely that the embryo itself is infected; not only do the young larvae sometimes die of the disease before emerging from the egg but the high occurrence of latent virus infections is against mere external contamination of the egg. Large insect populations may be infected with a virus which is passed from generation to generation without disease symptoms, until some stress such as overcrowding or other adverse conditions stimulates the latent infection into activity.

In addition to these two main methods of insect-virus spread, there may be occasional infections brought about by the activities of parasitic insects which possibly carry virus on the ovipositor. Again carnivorous insects may also help to spread the virus by eating virus-infected larvae and ejecting virus along with the faeces.

Plants

Here again, we are not concerned with the vectors of viruses, since these are dealt with in the next chapter. There are, however, several other means by which plant viruses can be spread.

The number which are transmitted through the seed is small, though not as small as was once thought. The reasons for this comparative rarity of seed transmission are not known, though various conjectures have been put forward, the most likely being the anatomical isolation of the embryo. Two authentic instances of seed transmission occur with the viruses of bean and lettuce mosaic.

In contrast, there is almost invariable transmission of a virus disease by means of the vegetative propagation of plants. All plant viruses which are systemic in their hosts are transmitted through vegetative organs such as tubers, bulbs, rhizomes, cuttings, etc. This makes the virus diseases of such crops as potatoes, strawberries, bulb plants, raspberries and hops of such great economic importance.

A few plant viruses which occur in high concentration in their hosts are transmitted by contact, that is by a diseased plant knocking against its neighbour in the wind. A good example of this mode of dispersal is given by the ubiquitous virus known as potato virus X which has no insect vector but occurs wherever the plant is grown. It spreads in the field by contact between both roots and haulms of X-diseased potato plants. Unlike most plant viruses, but similar to some animal viruses, potato virus X is carried on the boots and implements of farmworkers.

Another stable and highly infectious plant virus, which also has no insect vector, is that of the classical tobacco mosaic. This virus is carried on the hands and implements of workers in tomato houses and tobacco fields and retains its infectivity after the tobacco leaf is cured. This means that most brands of pipe tobacco and cigarettes retain the tobacco mosaic virus in viable form, and so act as a source of infection.

For many years an axiom has been laid down that very few plant viruses were transmitted through the soil; now we know that this axiom is profoundly untrue and that there is a large and increasing group of plant viruses that are soil-transmitted. A number of these are undoubtedly spread by soil-inhabiting organisms, and this point is discussed in chapter 7; but there are other viruses, notably the tobacco necrosis group, which are spread simply by mechanical contact between roots and virus. In these cases it is necessary for a slight wound to be made, the rupture of a root hair is sufficient, to enable the virus to enter as the virus is unable to enter an intact cell. This is analogous to the spread of fowl-pox virus, where it was necessary to add grit to the drinking-water of pigeons to make a slight abrasion before the virus in the water could infect the birds.

The methods of plant-virus dissemination which we have discussed are the natural means of spread. There are also artificial methods. One of these is inoculation, whereby infective sap is rubbed gently on the leaves of healthy susceptible plants; not all plant viruses are transmissible in this manner, although many are, particularly those of the mosaic group such as tobacco and cucumber mosaic viruses. All plant viruses which are present throughout their hosts (= systemic) are transmissible by grafting; this is a mainly artificial method of virus dissemination, though natural grafts do occur from time to time. Included in this is the action of the parasitic plant, *Cuscuta* sp., which by causing a natural graft between two plants can transmit a virus. In fact *Cuscuta* has now become a useful laboratory tool, because by making it possible to graft incompatible plants, viruses can be studied in new and unusual plant hosts.

Virus reservoirs and sources of infection

The questions of what happens to viruses in between epidemics, or where the viruses come from when they suddenly appear, are of great interest and importance. To paraphase an old song 'where does the 'flu go in the summer time?' We know, of course, that part of the answer lies in the existence of 'carriers' in which there is a latent virus infection, a phenomenon particularly common in plants and insects; that question is more fully discussed in chapter 9; here we are more interested in virus reservoirs existing in organisms of quite a different type from that in which the disease may eventually develop. One answer is to be found, unfortunately, in many of the pets and domestic animals with which man is liable to be associated.

We have only to look at the medical journals or even the daily press to find examples of virus infections in man brought about by contact with some domestic animal. These diseases are known as 'zoonoses' but this category includes, of course, disorders due to other agents besides viruses.

Suppose we consider first birds of the psittacine type, budgerigars and the various kinds of parrot. A fatal case of poliomyelitis in a child in which the infection was apparently acquired from a budgerigar was reported in *The Lancet*. Two budgerigars kept in the same cage developed an illness in which their legs were paralysed, and one died. While the survivor was recovering and was allowed out of its cage one day it took hold of the lip of a boy aged nine, and held on until removed by the boy's mother. A week later the boy developed bulbar poliomyelitis, which proved fatal four days later. Type I poliovirus was isolated from the boy and from the budgerigar. The conclusion is that the birds became infected with poliovirus and that the boy was infected by the bite from the surviving bird.

This happily seems a rare occurrence, but a much more common event is for persons to be infected with psittacosis from sick parrots or budgerigars. *The Times* recounts such a story from the medical press under the title of 'Death in the Parrot's Cage'. This was the case of the proprietress of a Midlands nursing home who bought a parrot and trained it to sit on her shoulder while she visited patients. A surgeon met her one day as she left a patient's room, noticed that the bird looked sick and insisted that it should have no contact with any of his patients.

Within a few days the bird was dead. Shortly afterwards the proprietress developed broncho-pneumonia which proved fatal. Her personal maid, who had shared the care of the parrot, had a similar illness which also proved fatal, and a third member of the staff had a milder illness from which she recovered.

The psittacosis virus, however, has a wide host-range and in consequence many other different kinds of birds are reservoirs of infection. It was reported by the Medical Officer of Health for Liverpool in 1956 that two cases of psittacosis occurred in one family during that year, and the evidence suggested that the infection was contracted from homing-pigeons kept in the patients' home. A subsequent survey, carried out by the Liverpool University department of veterinary pathology on wild and homing pigeons, showed that psittacosis virus was isolated from five of the 230 wild pigeons. In blood tests made from fifty homing pigeons, at least 18 per cent presented evidence of having had, at some time or other, infection with psittacosis virus. There is also evidence that London's pigeons, and possibly the starlings as well, harbour a strain of psittacosis virus, so that too close an acquaintance with the pigeons in Trafalgar Square would be better avoided.

It is thought that the domestic cat can transmit to man the virus of feline enteritis; this is a highly infectious disease from animal to animal and probably 70 per cent of all cats are infected at one time or another. According to a paper in the *Proceedings of the Royal Society of Medicine*, six cases of infective polyneuritis were directly related to an illness of the patient's cat. The most interesting of these concerned a farm labourer who developed the disease. While he was ill his cat was sent to be looked after by a woman who lived 40 miles away. A month later the cat sickened and died of the condition known as feline enteritis, and the woman was admitted to hospital with infective polyneuritis and died 10 days later.

In a recent survey in *The Times* of pets as a danger to man several reservoirs of virus infection are mentioned. Cat-scratch fever, the name of which explains itself, is a condition of fever with enlarged glands; its exact cause is not known, since efforts to isolate a virus have failed and the cat responsible does not show any signs of illness.

As regards possible animal reservoirs of influenza virus the situation is somewhat uncertain but suggestive. The virus of swine influenza, which has existed in the U.S.A. since the great pandemic of 1918–19, has now been shown to be due to a type A influenza virus. When the 1957 pandemic of the Asian strain swept over the world, the World Health Organization instituted an animal serum survey in twenty-five countries to find out first if this virus existed in an animal population before the pandemic, and secondly whether it would establish itself in any of the domestic animals known to be susceptible to influenza A virus. Definite conclusions have not yet been drawn but it is now known that in addition to swine, horses, fowls, and ducks have been incriminated as potential virus reservoirs.

One of the rickettsiae which are associated with arthropods causes the disease known as Q-fever. It was first described in Australia in 1935, but has since spread to fifty countries in all five continents. In the original investigation the arthropod concerned in the life-cycle was a tick and the other host an insectivorous marsupial known as a bandicoot. Now, however, certain other wild animals, particularly rodents, have been found naturally infected; birds, including house-martins and swallows, also harbour the rickettsiae, and it is an interesting question whether these are responsible for the spread of the disease round the world.

Finally, we come to the lowly cockroach, which is hardly a pet but is nevertheless intimately associated with man in many parts of the world. This tough and resilient insect has been incriminated by Drs Roth and Willis of the U.S. Army as a highly dangerous reservoir of human disease. They have identified eighteen species of cockroaches which can transmit infectious agents by contact or by actually biting man. Several of the commonest species have been captured repeatedly in sewers, cess-pools and septic tanks and have been found travelling into buildings.

Four strains of poliovirus have been found occurring naturally in cockroaches and, in addition, they harbour encephalomyelitis and yellow fever viruses.

In the next chapter in the discussion on the vectors of plant viruses we shall find that nematode worms are incriminated, and whilst it cannot be said that nematode worms are actual vectors of animal viruses they do play an important role as reservoirs of infection. The fascinating story of swine influenza in the U.S.A. has been elucidated by Shope. This disease prevails in epidemic form each year in the middle western United States, generally from October to December, and is largely absent from this region during the remainder of the year. Once the disease has started in a drove of pigs it is highly contagious and there is no difficulty in explaining its spread. The puzzling aspects were the origin of the virus causing the first case of the disease, and also the question of where the virus hid out during the long period between epizootics. An interesting fact about these outbreaks was that they developed simultaneously in a number of droves of pigs within the same area, thus giving the appearance of a miraculously rapid spread.

However, we know now that this apparently rapid spread was really not a spread at all but a simultaneous activation or stimulation of a masked virus infection already present in the pigs. The masked virus needs a 'stress factor' to stimulate it into virulence, and under natural circumstances this is usually a spell of adverse weather.

It was an apparent superstition held widely by farmers in the American middle west that first put Shope on the right track in his investigations. This was the belief that the earthworm had something to do with the outbreaks of swine influenza, and of course it is a well-known fact that pigs are fond of eating earthworms. Now the earthworm is frequently the host of a parasitic nematode worm which spends part of its life-cycle in that host and part in the lungs of the pig. The next step in the elucidation of the natural history of swine influenza was to see if the lungworms were playing an essential part in the story. A number of pigs were inoculated with the influenza virus and then killed on the third, fourth and fifth day respectively after the onset of the disease. Lungworms were taken from these pigs, mixed with pig faeces and placed in a barrel of soil; to this were added about 400 earthworms. Five weeks later some of the earthworms were examined and found to be infested with lungworms. Then came the crucial test: would pigs fed with these earthworms develop swine influenza? The test was carried out but the pigs remained healthy. Something was missing, and further investigation showed that a 'stress factor' was needed to stimulate the masked virus into action. We have seen that in Nature wet inclement weather has this effect, but in his original experiments Shope found that intramuscular injections with suspensions of the bacterium *Haemophilus influenza suis*, either living or killed, induced the development of influenza. In one such experiment, the selected pig was fed seventeen earthworms containing third-stage lungworms carrying masked swine influenza virus. Forty-three days after this feeding it was exposed outdoors to inclement weather for 18·5 hr. Four days later the animal came down with an illness typical of influenza, and swine influenza virus was demonstrated in its respiratory tract.

After this successful demonstration of the biology of swine influenza the question followed naturally as to whether another virus of pigs, causing what is known in the U.S.A. as hog cholera and in Great Britain as swine fever, did not behave in a similar manner. Shope's findings with hog cholera closely paralleled those with swine influenza. He found that the swine lungworm can serve as reservoir and intermediate host of cholera virus and can transmit it, usually in a masked, non-pathogenic and non-immunogenic form, to swine. Animals infected with this masked virus remain normal to all outward appearances, until they are subjected to some stress capable of provoking the masked virus into virulence. Once unmasked by provocation, the virus induces clinically characteristic hog cholera.

Shope has had some difficulty in finding a stress factor which would regularly provoke the masked cholera virus. It is interesting to find that the most successful stimulation was brought about by the intervention of still another worm, the so-called round worm, *Ascaris* sp.

In these fascinating studies there are of course gaps in the knowledge of the process of infection. It is not known, for example, whether the occult or masked viruses remain within the lungworm intermediate host or whether the lungworms act as genuine vectors of the virus and infect cells with the masked virus during their migration through the body of the animal.

This work on the biology of the viruses affecting pigs emphasizes the extreme versatility of viruses in general, and, above all, their rather alarming capacity to accommodate themselves to almost any type of living organism.

7
VECTORS OF VIRUSES
Animal viruses. Plant viruses

Many of the viruses affecting both plants and animals are dependent upon other organisms for their transport from diseased to healthy host. These *vectors*, as they are called, are mostly, but not entirely, insects, and they are of course parasitic in the sense that they must feed upon the animal or plant in question in order to transmit the virus.

It is only in a few cases that the relationship between virus and vector is a purely mechanical one. Often the vector itself may act as an intermediary host or reservoir for the virus and in these cases the connexion is an interesting one, since it involves the virus in a relationship with very different kinds of organisms; this is especially true with some of the viruses affecting plants.

Animal viruses

The first demonstration of such a thing as the vector of an animal virus was made in relation to yellow fever. It was in Cuba, in 1900, when this disease was rampant among American soldiers taking part in the Spanish-American war, that a local doctor, Carlos Finlay, proclaimed to an unheeding world that there was some connexion between mosquitoes and yellow fever. But it was Walter Reed, the head of an American army commission, who finally proved that the yellow fever virus was transmitted by the mosquito *Aëdes aegypti*.

Since those early days the number of viruses now known to have arthropod vectors is very large, at least seventy-seven having been described, and they are grouped together under the name of 'arbor-viruses', which does not, however, carry any classificatory significance.

Before we discuss some of the interesting facts relating to animal viruses and their vectors we should mention some cases where there is no relationship between the two, other than a purely passive carrying of the virus by the vector, in other words 'mechanical transmission'.

A good example of this is the spread of the rabbit myxoma virus in Australia by the mosquito, called by Fenner 'the flying pin'. Here, as far as we know, the virus is transmitted only because it contaminates the outside of the mosquito's stylets, and once the virus has been rubbed off the insect can no longer transmit until it has contaminated its stylets again by another feed on an infected rabbit. This, as we shall see later, is a very different matter from what takes place when there is a biological relationship between the virus and its arthropod vector.

Another case of a purely mechanical transmission is given by the virus of fowl-pox, which again is carried on the outside of the mouth-parts of mosquitoes and other blood-sucking insects which feed upon fowls. This virus, as we have already seen, is easily transmitted through any casual wound such as would be made by a piece of grit in drinking-water containing the virus.

In the spread by mosquitoes of such diseases as yellow fever or equine encephalomyelitis, one fact is suggestive and, indeed, signifies that here we have no simple mechanical transmission. This fact is the delay between the time of feeding by the insect on the infected host and the time when the insect itself becomes capable of infecting another susceptible host. This delay in the development of infective power on the part of the insect indicates that something is happening to the virus which it has ingested. Probably two factors are involved in this delay: the virus must first multiply within the insect until a sufficient quantity is produced for an infective dose and, secondly, the virus must pass from the lumen of the gut to the site where it can multiply and from there to the salivary glands for ejection in sufficient quantity to cause infection.

From the studies of several workers it seems as if the cells of the gut walls may play a significant part in determining the capacity of an insect as a vector by acting as a barrier to the movement of the virus from the gut to other sites in the body. Two American workers, Merrill and TenBroeck, were among the first to show that by puncturing the wall of the alimentary canal the mosquito *Aëdes aegypti* was enabled to transmit a strain of equine encephalomyelitis virus which it could not otherwise have done.

This peculiar significance of the gut-wall in relation to virus transmission will be mentioned again, when we are discussing the vectors of certain plant viruses.

In the mosquito species, which are normally the vectors of the yellow fever or encephalomyelitis viruses, the virus is carried for long periods and probably for the rest of the life of the insect. In these cases therefore the mosquito is a definite host of the virus and plays an essential part as a reservoir in the cycle of hosts, in which man is probably merely an incidental factor.

What evidence have we then that there is a biological relationship between the mosquito vector and the yellow fever or equine encephalomyelitis viruses? We have already seen that a mosquito carrying the virus of yellow fever does not become immediately capable of transmitting it, but that there is a delay whilst the virus is presumably multiplying inside the insect. There is, however, more direct evidence than this; the same two American workers, Merrill and TenBroeck, mentioned above, experimented with serial inoculations of mosquitoes. They ground up some mosquitoes known to be carrying the equine encephalomyelitis virus and inoculated this into some virus-free mosquitoes. These in turn were treated similarly and so on, in series. In this way it was shown that the virus persisted through a number of passages which involved too great a dilution unless the virus actually multiplied in the bodies of the mosquitoes.

For many of the arborviruses (=arthropod-borne) the vectors are not mosquitoes and in fact they are not insects at all but other kinds of arthropods; these are the mites and ticks which are concerned with the rickettsial type of disease agents, which for our purposes are regarded here as viruses. The rickettsiae are considered to have been originally microparasites of arthropods which have become adapted to propagation in man and the higher animals. With some of the rickettsiae a complicated cycle of hosts has been established in which there is a symbiotic relationship, one of mutual tolerance, without the development of disease symptoms except when an incidental host such as man becomes involved. A good example of this is scrub typhus or tsutsugamushi fever, which is transmitted by a trombiculid mite.

Of all the arborviruses it seems as if the rickettsiae were the only agents which are transmitted transovarially from generation to generation of the arthropod vectors through the successive stages of nymph, adult, egg, larvae, etc. The rickettsiae seem to have no ill-effect on the health of the mite—a different state of affairs from that existing with some of the insect-transmitted plant viruses we shall be discussing later.

The life-cycle of the vector of scrub typhus is rather a complicated one, as it is only in the larval stage that the mite becomes a blood-sucking parasite of a vertebrate host, which in this case is the field rat, the other stages being phytophagous. The rat thus becomes infected, but, like the mite which infected it, suffers no apparent ill-effects; the rickettsiae multiplying in the rat are picked up by other larval mites which happen to be parasitizing the rat, and so the cycle is completed. There is thus a symbiotic relationship between virus, vector and host, in which a mutual tolerance is achieved. It is only when man comes by chance into the picture that the clinical disease of scrub typhus develops. The disease is seldom transmitted from man to man because larval mites do not have the opportunity of feeding upon man whilst the rickettsiae are still available in the blood.

In studying the epidemiology of the arborviruses one important factor is the existence of a reservoir or 'maintaining' host for the virus.[1] The epidemiology of arborviruses can only be understood in relation to the vertebrate–invertebrate–virus cycle, which is known as a 'biocenose'. The survival of an arborvirus in a given area probably always depends on the stability of the entire biocenose, both vertebrate and invertebrate reservoir hosts being necessary. Often incidental hosts such as man are infected and an 'incidental biocenose' may develop which maintains the virus for a time, but its survival is not dependent on this. In the reservoir biocenose the host–vector–virus relationship has reached a balance in which neither vector nor host is adversely affected. This is illustrated by the monkey–mosquito reservoir of yellow fever.

For a long time a careful search was made for a reservoir host for the equine encephalomyelitis virus during inter-epidemic periods, and among the suspects was the egret. It is now thought that birds

[1] This subject has recently been discussed by Gordon Smith (*Brit. Med. Bull.* **15**, 1959) and much of the following information is derived from his interesting article.

and mosquitoes maintain the eastern strain of this virus in fresh-water swamps of the eastern U.S.A. On the other hand a search has recently been made for this strain of virus (EEE) in hibernating mosquitoes in Connecticut. During the winter of 1956–7 hibernating mosquitoes were collected from seven localities in Connecticut where there had been history of EEE activity. Two thousand five hundred and sixty-nine specimens representing six species were collected and tested for the presence of virus. Here again we see one of the practical applications of the tissue-culture techniques; in order to make the virus test the mosquitoes were ground up and inoculated into chick-embryo tissue cultures. Thirty-one inoculations were also made into infant mice, but in neither test was any virus isolated. The authors of this paper point out that although their studies produced no evidence of EEE virus overwintering in hibernating mosquitoes they do not eliminate the possibility of this occurring.

In the case of the western strain of this virus (WEE), however, positive results were obtained from one of thirty-three batches of *Culex tarsalis* collected during the winter in Colorado.

Research has revealed the existence of several interesting and unsuspected maintaining hosts. For example, the virus complex causing the Russian spring–summer encephalitis (RSS) passes the winter in hibernating hedgehogs. If these animals become infected just before hibernation, the virus circulates in them throughout that period without the development of antibodies; the virus of St Louis encephalitis apparently persists in a similar manner in bats.

We have seen in chapter 6 how cockroaches can become reservoirs of viruses, including poliovirus and yellow fever virus, and these are sometimes eaten by monkeys and could thus play a part as a maintaining host.

In these days of rapid communication between countries an ever-present danger exists in the potential transport, especially by air travel, of insects and other invertebrates. This danger is twofold: first, the insects may be carrying a virus which is not present in the country of arrival, the introduction of mosquitoes bearing yellow fever virus into India for example; and, secondly, the insects themselves may become established and act as new hosts for viruses already present. In consequence there are stringent regulations for the elimination of possible vectors of viruses in aircraft arriving from countries which are a potential source of danger.

There are also, of course, precautions to prevent the introduction of yellow fever to Asia by air travellers themselves during the incubation period.

Migrant birds are sometimes important as long-distance disseminators; the virus of St Louis encephalitis is enzootic in birds in the U.S.A. and has been isolated from birds in Trinidad. Similarly, birds are apparently responsible for the introduction of Murray Valley encephalitis virus to South Australia from the tropical north by migration in stages, with spread between birds at each stage.

There is a good deal of circumstantial evidence that the virus of foot-and-mouth disease may be carried by migrating starlings from the continent of Europe into the British Isles. If this is so, it is a case only of mechanical contamination of the feet or feathers of the birds which are in no sense a host of the virus.

It sometimes happens in a reservoir biocenose that a particular vertebrate host may provide such large amounts of virus that infection overflows into neighbouring incidental vertebrates among which man is usually included. This type of virus reservoir is known as an 'amplifier'. A case of this occurs in Japan where epidemics of Japanese encephalitis are heralded by the intense spread of the virus among nestlings in large colonies of herons. The explanation probably is that the bird-biting mosquitoes responsible for the spread of the virus bite man and horses comparatively seldom and so, when a very high proportion of mosquitoes is infected, the spread is much greater. The horse itself and other large vertebrates may also act themselves an amplifiers between birds and man.

Plant viruses

The vectors of plant viruses belong mostly, but with a number of interesting exceptions, to the insect order Hemiptera which includes all those insects with sucking mouthparts; these feed by sucking the plant-sap and do not, like the biting insects, ingest any plant tissue.

The mouthparts of this type of insect consist of two pairs of fine needles, the mandibles and the maxillae, which are contained in a supporting structure, the 'beak' or labium. On the inner faces of the mandibles and maxillae are two grooves which are brought together to form two fine canals. Down one of these canals passes the saliva of the insect and up the other passes the sap of the host plant together with any virus present in the sap. When the insect feeds it presses on the leaf-surface with the labium; this, however, does not enter the plant but, being fitted with an open grove on the upper surface, it bends and becomes foreshortened, so allowing the stylets to pass down into the plant tissue. For a long time it was thought that the plant-sap was drawn up into the insect by means of a pharyngeal pump situated in the head. Now, however, some doubt has been felt about this, and it is suggested that the turgidity of the plant cells themselves forces the sap up the canal in the mouthparts. This suggestion arises from the fact that if the head of an insect is removed, whilst sucking, the sap continues to flow through the stylets left *in situ* in the leaf.

The relationships between plant viruses and their insect vectors can, as in the case of the animal viruses, be divided into two classes, that in which a biological relationship exists between the two and that in which there is apparently no such relationship. The situation as regards mechanical transmission, however, is by no means as clear as with the animal viruses, and there is no case analogous to the myxoma virus and the 'flying pin'.

It will be clearer, perhaps, if we take two of the large groups of insect vectors and deal with them separately; these are the aphids, or greenflies, and the Jassids or leafhoppers.

It is probably true that a majority of the plant viruses are transmitted by aphids, and one particular species, *Myzus persicae*, the peach and potato aphid, is responsible for the spread of at least fifty separate viruses in different parts of the world.

The aphid-transmitted viruses fall into two groups, the 'non-persistent' and 'persistent' viruses; these terms refer to the length of time the virus is retained by the aphid without recourse to a further source of infection. Non-persistent viruses are rapidly lost by the insect usually after one feed on a susceptible plant, persistent viruses on the other hand are retained for long periods by the aphid, and frequently for the rest of its life. There is, however, no hard and fast division between the two types, and viruses occur which have some characteristics of both.

Our knowledge of the exact biological relationships between aphids and the viruses they transmit is rather unsatisfactory and in spite of much research the situation is by no means clear.

Probably the non-persistent aphid-borne viruses are mechanically transmitted in the sense that there is no biological relationship between virus and aphid. It is indeed possible for the whole procedure of transferring the virus from the infected to healthy plant to be completed in under one minute. On the other hand not every species of aphid can transmit the same virus, and starving the insect before allowing it to feed on a virus source increases its efficiency as a vector. Various suggestions have been put forward to explain these anomalies; for example, the virus may be adsorbed on to the outside of the stylets, particularly in pockets formed by minute hooks which are present in some species but not in others. Another possibility is the contamination of the inner lumen of the stylets with virus which causes infection until the tubes are cleared by the aphid in process of feeding. As for the phenomenon of increase in vector efficiency by starving the insect before feeding it on a virus source, it has been suggested by Bradley that it is the result of a change in the behaviour of the aphid invoked by the starving. In other words instead of settling down in one spot and inserting its stylets deeply into the phloem of the leaf, the aphid tends to wander over the leaf-surface making short punctures into the epidermis only. This behaviour favours the spread of non-persistent viruses which occur mainly in epidermal cells and not in the phloem.

These facts are perhaps sufficient to indicate that whilst the aphid-transmission of non-persistent viruses may be mechanical it is not so straightforward as the 'flying pin' method with the rabbit myxoma virus.

Now as to the persistent viruses, here again opinion is divided on the exact nature of the relationship between this type of virus and its aphid vector. The main differences between this and the foregoing are that the virus is retained for long periods after only one virus-acquisition feed, starving has no effect on efficiency, and there is frequently a delay in the development of infective power in the insect after it has fed. In other words there may be a waiting period of several hours or even days in some cases after feeding on a source of virus before an aphid becomes infective to a susceptible plant. Another point of difference from the non-persistent viruses is that the aphid vector of a persistent virus must feed for some time, usually several hours, before it can pick up the virus. Part of the explanation of this may be that most persistent viruses are located in the phloem and it takes the aphid some time to penetrate thus far into the tissue.

The question of most interest is whether any persistent viruses actually multiply inside the bodies of the aphid vectors. The evidence for such multiplication is rather meagre and what there is centres around the persistent virus of potato leaf-roll and the aphid *Myzus persicae*. After feeding for a day or two on a source of leaf-roll virus, the aphids retain infectivity for as long as a week, even after feeding on a host such as the cabbage plant, which is immune to the leaf-roll virus.

The most direct evidence of multiplication is afforded by some recent work in Holland where the technique of serial transmissions from aphid to aphid was employed. This method was used, as we have already seen, to demonstrate multiplication of equine encephalomyelitis virus in the mosquito. As may be imagined it is a difficult matter to inject virus into aphids without killing them and the Dutch workers designed an ingenious micro-injection apparatus. With this they were able to extract a drop of fluid from a virus-bearing (viruliferous) aphid and inject it into a virus-free aphid; the injected aphid then became in its turn viruliferous. After a number of successful serial transmissions, it was considered that the virus must have multiplied in the aphid, as otherwise the dilutions involved would be too great for the survival of the virus. On the other hand, some circumstantial evidence does not support this thesis; aphids do not apparently retain the leaf-roll virus indefinitely, infectivity seems to diminish with the lapse of time. In addition, experiments have shown that high temperatures reduce the efficiency of the aphids as vectors of the leaf-roll virus, and efficiency is not restored when the aphids are returned to lower temperatures. This is the contrary to what would be expected if the virus was multiplying in the insect.

When we come to consider the other large group of sucking

insects, the leafhoppers, and the viruses they transmit, we find a great deal more positive evidence on the biological relationships between insect and virus. Plant viruses which have leafhopper vectors have several points in common; with rare exceptions they are not transmitted mechanically but are dependent upon one or more species of leafhopper, and they are not transmitted by any other kind of insect. The symptoms on the plant are of the same general type, consisting mostly of malformations of flowers and leaves and the absence of a leaf-mottling characteristic of many aphid-borne virus diseases. Leafhopper-borne viruses are always persistent and are usually carried by the insect for the rest of its life. There is a delay in the development of infective power in the leafhopper after it has fed on a source of virus, just as we saw in the case of mosquitoes and yellow fever virus. We can also find another instance of analogous behaviour; it will be recalled that the mosquito *Aëdes aegypti* can only transmit the eastern strain of equine encephalomyelitis virus (EEE) if a puncture is made in the wall of the alimentary canal. In the plant-virus disease known as maize streak transmitted by a leafhopper, rejoicing in the name of *Balclutha mbila*, Storey has shown that there occur two morphologically indistinguishable races of the insect, one of which can, and the other cannot, transmit the virus. However, if a puncture is made in the wall of the alimentary canal of those insects unable to transmit they become at once potential vectors of the virus. Here then we seem to have some property in the wall of the gut which prevents the passage of the virus through it. These two examples are very similar, the difference being that with the horse and the mosquito there are two strains of virus and one strain of the insect, and in the streak disease of maize one strain of virus and two strains of the insect are involved.

In discussing the vectors of animal viruses we have seen that no virus, so far as is known, is transmitted transovarially in the mosquitoes or other insect vectors, although hereditary transmission is known to occur in the tick-transmitted rickettsiae. In the leafhopper-borne plant viruses, however, the virus is transmitted via the egg in several cases. The first to demonstrate this was a Japanese, Fukushi, working with the dwarf disease of rice and its leafhopper vector, *Nephotettix apicalis* Motsch. He showed that the virus was transmitted from an infective parent insect to the offspring, but only through the female parent. Moreover the progeny from such an infected parent did not itself become infective until after a period of nine days from the date of hatching. Fukushi also showed that the virus could be passed through six generations involving eighty-two infective leaf-hoppers and all derived from a single virus-bearing female without access to a further source of virus. More recently Black, in America, has carried out similar experiments with the virus of clover club-leaf which is transmitted through the egg of the leafhopper vector, *Agalliopsis novella* Say. From a pair of viruliferous leafhoppers the breeding was carried out through twenty-one generations over a period of five years. The insects were fed throughout on virus-immune lucerne plants without loss of infectivity.

These two cases of virus inheritance, whilst of interest in themselves, are also, of course, proof of the multiplication of plant viruses in their insect vectors. This is further evidence of the versatility of viruses in being able to propagate themselves in such different kinds of host and is an important point in our study of the relationships between viruses and their vectors.

More evidence, both direct and circumstantial, can be offered on this subject, and some selected experiments are given here because of their interest and importance.

Serial transmissions of virus from leafhopper to leafhopper, a technique we have met already, have been carried out by Maramorosch and others in the U.S.A., and these also are proof of multiplication of virus because otherwise the dilution involved would be too great. An earlier and more indirect experiment on these lines and one which aroused considerable controversy at the time was carried out by Kunkel, another American worker, who studied the disease of aster yellows and its leafhopper vector *Macrosteles fascifrons* Stal. He exposed viruliferous leafhoppers to a temperature of 32° C for varying periods and studied the effect of this on their ability to transmit the virus. He found that the high temperature deprived the insects of their infectivity for a period, and that the length of this period was dependent on the length of time of their exposure to the high temperature. In other words leafhoppers kept at 32° C for one day regained their infectivity in a few hours; if they were exposed for several days it required two days for them to regain infectivity, and exposure for 12 days resulted in permanent loss of infectivity. Kunkel interpreted these results as indicating that the amount of virus in the insect vector was reduced by the high temperature to a point below infectivity level, and a period of time was therefore necessary for the virus to multiply up again to reach the necessary concentration for infection. Long exposure to heat apparently inactivated all the virus in the insect which therefore lost infectivity until fed once more on a source of virus. In 1937 when these results were published a good deal of scepticism on their interpretation was felt by some virologists but now, in view of the mass of other evidence available, they are accepted at their face value.

Kunkel has carried out another interesting experiment on the transmission of aster yellows virus which also bears on the question of virus multiplication in the vector. To explain this it will be necessary to digress for a moment. Kunkel's experiment is concerned with the question of acquired immunity; as is well known, animals which have suffered from a virus disease develop antibodies against that particular virus, conferring an immunity for a period which varies with the virus. Plants having no antibodies develop no such acquired immunity; their only kind of acquired immunity is of the non-sterile type. In other words a plant already infected with a given virus is sometimes, but not always, resistant to infection with another strain of the same virus. There are two such strains of aster yellows virus, known respectively as the New York and California strains. Since the symptoms induced by these two viruses in most plants are so similar it was a long time before it could be determined whether they did in fact cross-immunize. However, Kunkel eventually found two plants which reacted differently to the two strains and by means of these he was able to show that there was a cross-immunity between the two. The transmission experiment was carried out as follows: leafhoppers were fed first on a plant infected with New York aster yellows, then transferred to a plant infected with California aster yellows, and finally to the healthy test plants. Next the experiment was repeated except that this time the order of feeding on the virus sources was reversed. The results were rather unexpected; in every case when the insect was fed successively on the two strains, it transmitted only that on which it fed first. Such a phenomenon clearly shows some biological relationship between insect and virus; it may be that certain cells, possibly of the fatbody, were occupied by the strain of virus first imbibed, and this precluded, not necessarily the entrance, but probably the multiplication of the second virus strain.

This experiment brings up a question which has long intrigued virologists; is the vector of a virus in any way harmed by the virus it transmits? There seems to be no instance, so far as the small viruses attacking animals are concerned, of the mosquitoes or other insect vectors being harmed in this way; mosquitoes carrying the virus of yellow fever seem to live as long as those which are virus free. There is of course the well-known case of typhus fever in which the louse vector is invariably killed by the rickettsiae, but this seems to be the only instance.

The situation is rather different with the plant viruses and evidence is accumulating that, in the case of one or two leafhoppers at least, there is a definitely deleterious effect on the vector of the virus it transmits. For example, Jensen, in the U.S.A., has shown that leafhoppers carrying a virus from peach trees have a much shorter life than the control leafhoppers. Perhaps the most unequivocal evidence of a plant virus harming its insect vector comes from the work of Watson and Sinha in England. Working with the virus of wheat striate mosaic, they have shown that infective females of the planthopper *Delphacodes pellucida* Fabr. which fed on infected plants as nymphs had 40 per cent fewer progeny than did those on healthy plants. Furthermore, the virus seemed pathogenic to the embryos, some of which died at a comparatively late stage of their development. Incidentally, this is another case in which the efficiency of the vector is increased by making a puncture in the wall of the alimentary canal.

We mentioned at the beginning of this discussion on the vectors of plant viruses that there were some interesting cases of transmission by vectors other than the Hemiptera. One of these concerns the virus of tomato-spotted wilt which is the only virus so far known to be carried by a thrips (Thysanoptera). This is a minute linear insect which does not suck the sap but feeds with a kind of pick-axe movement with its single mandible. The main point of interest about this vector is that it cannot acquire the virus in its adult stage; it can transmit provided that it has fed on a source of virus when a larva but it cannot pick up the virus *de novo* as an adult. Here again we may be up against the problem of the permeability of the gut-wall and a possible difference in the adult and larval stages.

We have not yet mentioned insects with biting mouthparts and the part they play in the transmission of plant viruses. The commonest plant-feeders of this type are caterpillars and beetles which actually ingest the leaf tissue, and there are several viruses which are spread by beetles. In the discussion of the mechanism of virus spread by sucking insects the saliva was shown to play an important part; it is interesting therefore to find that the absence of saliva in biting insects is the essential fact in virus transmission. The virus of turnip yellow mosaic is spread in the field by one or more species of flea beetles, but experimentally it can be transmitted by any type of beetle which will feed on the necessary susceptible plants. The explanation seems to be as follows; since beetles have no salivary glands it is necessary for them to regurgitate part of the contents of the foregut whilst eating; this apparently helps in the process of digestion. Regurgitation brings into contact with the leaf any infective matter previously eaten and this during the process of mastication is inoculated to a healthy susceptible plant. Virus transmission of this type is not confined to beetles and the virus of turnip yellow mosaic is spread by grasshoppers and earwigs which feed in a similar manner to beetles. The virus, however, is not spread by caterpillars which are also biting insects but which do not regurgitate whilst feeding.

Outside the Insecta but still in the Arthropoda, mites and ticks are concerned in the spread of the animal viruses, mainly rickettsiae; in plant-virus transmission we come across the mites again but not the ticks. There are at least four plant viruses which are spread by phytophagous mites, those of black currant reversion, fig mosaic, peach mosaic and wheat streak mosaic. Not much is known about the relationships between these viruses and their mite vectors except in the case of wheat streak mosaic; it has been shown by Slykhuis that when the different developmental stages of the mite were colonized on the diseased wheat the nymphs could acquire the virus but the adults could not. We have already met this phenomenon in the transmission of tomato spotted wilt by thrips.

In the preceding chapter an account was given of the association of nematode worms with the infection cycle of swine influenza. These worms are also involved in the spread of plant viruses and do in fact play a more active part since they are the actual vectors of a number of viruses. The first plant virus that was shown to have a nematode vector was that of the fanleaf disease of grapes in the U.S.A., and recently similar vectors have been found for one or two viruses affecting tomatoes and other plants in the British Isles which were known to be soil-transmitted. The nematodes in question seem to belong to the genus *Xiphinema*, but there is very little information yet available on the relationship between plant virus and nematode worm.

A comparison of the vectors of animal and plant viruses respectively and their relationships reveals some differences, but more similarities. Transovarial transmission of animal viruses occurs in mites and ticks but not so far as known in insect vectors, whereas this type of spread is thought to be confined to the insect vectors, leafhoppers, in the case of the plant viruses. Adverse effects on the vector of the virus transmitted have been observed in one or two cases of the leaf-hoppers transmitting plant viruses but not in the vectors of animal-viruses with the exception of the typhus rickettsiae in the louse.

The points of resemblance include multiplication of the transmitted virus in the vector with an incubation period, as in mosquitoes and leafhoppers. Both biological and mechanical transmission occur in the two types of viruses, and in certain cases the same question of the impermeability of the gut to some viruses arises. There is the same heterogeneous collection of arthropod and other vectors which includes biting and sucking insects, mites and nematode worms. Ticks, of course, are not concerned with the transmission of plant viruses since they are blood-sucking arthropods.

All these facts serve to emphasize not only the versatility of viruses in their capacity to adapt themselves to many different kinds of organisms, but also the similarities between plant and animal viruses. This is still further emphasized by the ability of a plant virus to multiply in an animal vector and it raises the question as to whether the leafhopper-transmitted viruses may not have started as insect diseases which later became adapted to plants.

8

VIRUSES AFFECTING INVERTEBRATE ANIMALS

Arthropods. Nematode Worms

Arthropods

Until recently the insects were the only invertebrate animals in which virus diseases had been observed, but now viruses have been isolated from other arthropods, the red spiders or spider-mites, many of which are serious pests of fruit trees, tomatoes and other crops.

This rather uneven distribution of viruses among the various branches of the animal kingdom is more apparent than real and is probably due to our lack of knowledge. The same apparently uneven distribution of viruses can be seen among the different orders of insects, but with increased investigation viruses are now being discovered among insect orders which had previously been regarded as free from virus attack.

We will commence this chapter with a short general account of the various types of viruses so far recorded from insects.

Insect viruses

One of the outstanding characteristics of the insect viruses is the enclosure of the virus particles in membranes, capsules and crystals. The purpose of these 'intracellular inclusions', as they are called, is rather obscure; they may be some kind of defence mechanism in the insect, but if so it is a singularly ineffective one since once the inclusions are found the insect almost invariably dies. On the other hand these inclusions, particularly the polyhedra, are an admirable protective covering for the otherwise rather unstable virus, and furthermore they assist the dispersal of the virus and its access to new hosts.

There appear to be three main groups of insect viruses; first there are the diseases characterized by the presence in the tissues of many-sided crystals or polyhedra; these are usually known as 'polyhedroses'. Second, there are the diseases characterized by the presence of granules; these are known as 'granuloses'. The third type has no intracellular inclusions, but the virus is free in the tissues as in the virus diseases of plants and the higher animals.

The polyhedroses can be subdivided into two main groups, the nuclear and the cytoplasmic diseases; the viruses concerned in these two types of disease are quite distinct and so are the symptoms produced. As the names imply, the site of virus multiplication is respectively the cell nucleus and cytoplasm.

Nuclear polyhedroses. Of all branches of virology the study of the insect viruses has been most neglected, and this is a pity because they are of great scientific and practical importance. During the last decade, however, more intensive investigation has been made of these viruses, and an interesting picture is now beginning to emerge. For many years almost the only type of insect-virus disease known was the so-called wilt disease; this attacked the caterpillars of lepidoptera, butterflies and moths, and we know it now as a nuclear polyhedrosis; the classic example was 'jaundice' of the silkworm to which we referred in chapter 2.

The caterpillar becomes infected by eating food contaminated with polyhedral crystals; these pass down into the alimentary canal where they are dissolved by the alkaline secretions of the gut. The liberated virus then makes its way to the blood and to the skin, where it begins to multiply. Sections made through the tissues of a caterpillar in the early stages of infection and observed on the optical microscope show the polyhedral crystals beginning to form in the cell nuclei. As more and more crystals are formed, the nuclei become bigger and bigger until finally they burst, liberating thousands of polyhedra into the cytoplasm. As the disease progresses the cells burst and then the skin of the caterpillar, which by this time has become extremely fragile, also ruptures and millions of polyhedra are liberated. Indeed a striking aspect of this type of disease is that the whole body contents become liquefied.

We said that the polyhedra, on entering the gut of the larva, are dissolved by the alkaline secretions and the virus is liberated. This process can be carried out in the laboratory and observed under the electron microscope; weak sodium carbonate dissolves the crystal, and the virus rods, thus liberated, can be seen, sometimes singly, sometimes in bundles. A characteristic of the nuclear polyhedra is the presence of a thin membrane which encloses the crystal; this membrane does not dissolve in the sodium carbonate but remains behind with virus rods frequently still occluded (plate XIV, A). It is thought that this membrane is probably formed by the hardening of the outer layer of the protein crystal.

Nuclear polyhedroses have been described from many caterpillars of the lepidoptera and from the larvae of the sawflies (Hymenoptera), but so far only one virus giving rise to nuclear polyhedra has been observed in the larvae of the true flies (Diptera). This particular virus is of considerable scientific interest, and a short account of it is given here. It attacks the larva known as a 'leatherjacket' which develops into the familiar crane-fly or 'daddy-long-legs'. The site of virus multiplication is the blood, and a kind of leukaemia is caused in which the larva becomes filled with great quantities of blood cells. The polyhedra are formed in the blood cells at first round the periphery of the cell nucleus, but later, on the rupture of the cell, they are liberated into the haemocele. Strictly speaking these 'polyhedra' are not many-sided but are shaped more like a half-moon or the segment of an orange. When treated with dilute sodium carbonate their behaviour is most peculiar and is quite different from that of the polyhedra of other nuclear polyhedroses. Instead of dissolving and liberating the virus rods contained within, the polyhedra are pulled out or elongated into a worm-like shape; when the solution is neutralized again the polyhedra resume their normal shape, and this procedure can be repeated indefinitely provided the solution never becomes too alkaline; if it does, elasticity is lost and the polyhedra remain in the extended condition. Since this treatment was useless for showing the presence of virus in the polyhedra another method had to be used. Ultrathin sections were cut of the polyhedra and viewed under the electron microscope; these showed the presence of large numbers of rod-shaped virus particles. Now there are two points of interest about these particular polyhedra. One is the apparent orientation of the particles inside the 'crystal' into a regular order as compared with the haphazard arrangement of the virus rods in the polyhedra of Lepidoptera; and the other is the difference in the texture of the polyhedra from the fly larva and that from the Lepidoptera, and this is the absence in the former of a crystalline lattice at high magnification. Plate XII, which is an ultrathin section through an infected blood cell from a larva of *Tipula paludosa*, shows both these points; it also suggests why the virus particles are regularly orientated. They can be seen adsorbed in rows along the inner face of the inclusion body and, as they adsorb, a layer of polyhedral material (probably protein) covers them. The virus particles are therefore orientated in regular rows rather like the growth-rings in a tree.

The cytoplasmic polyhedroses. The existence of this type of disease was discovered quite by chance, and it is now known to be extremely common, probably more so than the nuclear polyhedroses. The routine method of testing for the presence of the nuclear polyhedra in an insect is by making a smear of the insect's blood and tissues on a microscope-slide, fixing it by heating over a bunsen-flame, and staining it with Giemsa's solution; this process shows up the nuclear polyhedra by leaving them unstained against a stained background. However, whilst carrying out this procedure some years ago the writer noticed that a number of the polyhedra had also stained. When this type of polyhedra was treated with weak sodium carbonate and observed under the electron microscope, a quite different state of affairs from that obtaining with the nuclear polyhedroses was revealed. Instead of dissolving completely and leaving behind a thin membrane containing the virus rods (plate XIV, A), the crystals had dissolved only partially, and the residue showed numerous round holes (plate XIV, B). Subsequent investigation showed that these holes were the sockets of a virus which was spherical in shape and not rod-like. Unfortunately these spherical viruses are destroyed by weak alkali before the polyhedra completely dissolve, and this makes it rather difficult to extract the virus from the polyhedral crystal. However, improvements in technique have now made it possible to obtain the virus in a pure state, and they are found to be spherical, or near spherical, as was first suggested by the shape of the sockets remaining in the partially dissolved polyhedra.

The granuloses. The second large group of insect-virus diseases, the granuloses, appears to infect only the larvae of the lepidoptera, the butterflies and moths. When a smear is made of the tissues of a larva infected with this kind of disease and observed under an oil-immersion lens, huge numbers of very minute granules can be seen forming the background of the field and no polyhedra are present. These granules are just at the limit of the resolution of the optical microscope, and are shown to be crystals under the electron microscope. When treated with sodium carbonate in the manner described for the polyhedra the granule partially dissolves and collapses, revealing a slightly curved rod-shaped body within. This was at first mistaken for the virus itself, but it is actually another capsule containing the rod-shaped virus particle. As a rule there is only one virus rod inside the capsule though occasionally two are present.

The granulosis disease resembles the nuclear polyhedrosis; the skin is similarly attacked and the body contents liquefied. Caterpillars killed by the disease remain attached to the food-plant and hang downwards frequently in the shape of an inverted V. The skin becomes very fragile and easily ruptures, liberating vast quantities of granules contained in the liquefied body contents. This, of course, facilitates the spread of the disease and plays an important part in the control of insect pests (see chapter 12).

Viruses without intracellular inclusions. This type of virus occurs free in the tissues of the insect as do virus infections of plants and the higher animals, and no polyhedra or granules are associated with the virus.

So far only two viruses of this kind have been described. One is very small and has been isolated from a caterpillar of the type, called in America 'army worms', because they move from one area to another in large numbers. This virus has been observed under the electron microscope and appears to be about 25 mμ in diameter. Affected caterpillars are swollen and darker in appearance than normal insects. The cuticle has a waxy appearance and the middle part of the insect is slightly enlarged; there is no liquefaction or disintegration of the body characteristic of the nuclear polyhedroses or granuloses.

The other 'free' virus is the *Tipula* iridescent virus (TIV) which attacks the larva of the crane-fly or 'daddy-long-legs' *Tipula paludosa*.

We have seen earlier some of the interesting properties of this remarkable virus. It occurs in such high concentration in the living insect that it crystallizes spontaneously, and these micro-crystals are peculiar in exhibiting a blue or violet iridescence. Affected insects can therefore be recognized immediately by the bright blue colour showing through the skin.

It has long been an axiom that insect viruses are extremely specific in their action and will not infect even closely related species. There is now a good deal of evidence that this is an erroneous idea and that the specificity of insect viruses is not so rigid as was originally supposed. It is, however, rather a difficult point to prove because of the complication of latent virus infections which are constantly cropping up and clouding the issue (see chapter 9).

The discovery of the *Tipula* iridescent virus gave an opportunity to test specificity without the added complication of latent virus infections. This was made possible by the unusual characteristics of TIV which enabled it to be used as a marker. The pentagonal or hexagonal shape of the virus, its large size and extreme uniformity of shape, the enormous quantity produced in each insect, and, most characteristic of all, the blue or violet iridescence of the infected larvae made identity certain, since there is no other insect virus known at present which has these properties.

Cross-inoculation studies soon showed that TIV had a very wide host-range, and would not only infect other kinds of fly larvae but would also attack insects of quite different orders such as the caterpillars of the Lepidoptera and the larvae of beetles. This is the first known example of an insect virus being transmitted from one order to another.

Some examples of these cross-transmissions of TIV are of interest here. The first hint that the virus was not specific in its choice of hosts was given by the appearance of some minute larvae of the so-called 'fungus' gnats, Mycetophilids, which occurred in the same containers with infected *Tipula* larvae. Among the normal pale-coloured larvae were a number which stood out from their fellows by their bright blue or violet iridescence. When sections of these larvae were examined under the electron microscope the presence of large quantities of virus in the skin was confirmed and the blue colour was, of course, due to the optical properties of the microcrystals formed by the virus in the living insect. After this discovery attempts were made to infect other species of flies and the virus was successfully transmitted to the larvae of St Mark's fly, *Bibio marci*, and of the bluebottle fly, *Calliphora vomitoria*.

It was next found possible to infect many different kinds of caterpillars, and one of these, the larva of the large white butterfly, *Pieris brassicae*, was used in the mass-production of TIV. Since this caterpillar is very easy to breed there was little difficulty in producing a gram or more of pure TIV. When spun down in the centrifuge tube a brilliant iridescent pellet is produced, rivalling in beauty a large opal.

Further experimenting showed that the larvae of some beetles were also susceptible to infection, among them the mealworm, the wireworm and the grubs of the chafer beetle. The appearance of the latter was much improved by the striking blue iridescence.

It must, of course, be understood that TIV is an exceptional virus in many ways and that the other insect viruses do not have the same remarkable properties.

We have mentioned earlier that viruses appear to be very unevenly distributed among the various orders of insects, though this of course may be due to lack of knowledge. At the present time, apart from the experimental infections just described, viruses have only been definitely recorded from the Lepidoptera, Neuroptera, Hymenoptera and Diptera. Although there appear to be infectious disorders of the aphids or greenfly, which belong to the Hemiptera, no authentic virus has yet been isolated from them. The same applies to the Orthoptera and the Coleoptera, although a rickettsial disease has been recorded from the larvae of the cockchafer, *Melolontha melolontha*. The rickettsiae, like viruses, are obligate parasites; they are much larger than most viruses, but they also cannot be cultivated in a cell-free medium.

Arachnida viruses

The red spiders, or spider-mites, which are also arthropods but which belong to the order Arachnida, constitute a large group among which are many serious pests of crops. As we have seen in chapter 7, some mites belonging to this order act as vectors of plant viruses just as many insects do, and also, like insects, they have been found to suffer from virus diseases of their own.

One of these red spiders occurs in California and has long been recognized as a serious pest of citrus fruits. Not long ago it was noticed that some of these mites were walking in a peculiar way with very stiff legs and appeared to be partly paralysed; they also seemed to be suffering from diarrhoea. Furthermore the disease was spreading and was undoubtedly very infectious.

By the extraction and centrifugation of large numbers of diseased mites a virus was finally isolated and observed on the electron microscope. It is a very small virus measuring about 35 mμ in diameter and is an icosahedron in shape.

What is almost certainly a very similar virus has now been found, also in the U.S.A., attacking another species of red spider, the one that is such a pest of apple trees in Europe. There is thus some hope that it may eventually be introduced into the United Kingdom and used to control the fruit-tree red spider (see chapter 12).

Finally, what may be a virus of quite a different type is a 'virus-like factor' which is transmitted by the egg and which affects the number of setae or bristles on the forelegs of a species of spider-mite. These setae, which are sensory, occur in all males and some females; the daughters of females without setae also lack them, but if the female parent bears the setae, the daughters also bear them. Experimental breeding suggests that a suppressing agent in the non-seta stock prevents the development of setae and that this agent is transmitted transovarially.

It is possible that this may be a similar type of virus-like agent to the kappa-particle in *Paramecium* or the carbon-dioxide sensitivity of *Drosophila*, both of which are transmitted to the offspring.

Nematode worms

We come now to quite another kind of animal, the nematode worms; these are mostly very small worms, some of which are parasitic in plants and animals whilst others are free-living in the soil.

In chapter 7 we read about the part played by nematodes as the vectors of several plant viruses, one affecting the vine and others from cereals, and about the important part played by the pig lung-worm in the infection cycle of swine influenza virus. Now we find them as victims of a virus disease of their own. Workers at the University of Nebraska, U.S.A., who were studying a gall-forming nematode called the Southern Root Knot Nematode, noticed that all the larval nematodes in one batch were extremely sluggish and failed to make their usual galls in the roots of tomato plants. By mixing some of these sluggish nematodes with a large number of normal nematode eggs, it was discovered that the disease was infectious. Larvae, hatching from these eggs, were slow and jerky in their movements and the posterior position of their bodies appeared to lose power of locomotion first; many died soon after emerging. To test whether a virus was causing the disease a suspension of the affected larvae mixed with some bacteria was passed through a Seitz filter; none of the accompanying bacteria passed the filter but the filtrate was still highly infectious to healthy nematode larvae.

This observation is of some interest as being the first recorded virus disease from worms; it may also be important as a possible agent in the biological control of a serious economic pest.

9

LATENT VIRUS INFECTIONS

Detection. Stimulation

One of the most interesting problems of virology is how it is that an organism can be infected with a virus without showing any ill-effect; in other words there seems to be a kind of mutual tolerance between the organism and the virus. This phenomenon is extremely common in virus infections, and there are so many different types of *latency*, as it has been called, that the situation has become somewhat confused. Before we discuss the matter, therefore, it may be as well to clarify the position with a few definitions.

At a symposium held in Madison, Wisconsin, in 1957 a fair amount of agreement was reached among virus workers on what was really meant by a 'latent virus'. The following definitions were agreed upon: an *inapparent infection* is one which shows no outward sign of its presence. Where the inapparent infection is chronic and where there is a virus–host equilibrium the condition is called a *latent infection*. An *occult virus* is one for which the virus particles cannot be detected and the actual state of the virus cannot be ascertained. The word occult was preferred to 'masked' since that word has been used with several different meanings.

Then, as we have seen, some viruses are known to have developmental stages; for these the names *provirus*, *vegetative virus* and *infective virus* were selected. Definitions were also put forward to describe the effect of viruses on the cells of the host: where the effect is slight such viruses are said to be moderate, where the cell is killed they are *cytocidal*, whilst viruses intermediate in effect are called *sub-moderate*.

The condition of latency can be divided arbitrarily into four types. First there are viruses that cause evident symptoms at the time of infection; these symptoms disappear but may recur at intervals as the consequence of some stimulus. An example of this type of latency is herpes simplex in man.

Secondly, there are viruses that also initially cause a disease but the host recovers and, although it still contains the virus, never again shows external evidence of its presence. This category includes the virus of lymphocytic choriomeningitis in mice, which has been called by Andrewes an 'indigenous virus'. After the mouse has recovered from the infection the virus persists in the blood and, in the case of female mice, infection passes to the young in the uterus. After two years the virus has become so mild that its presence can only be demonstrated by inoculation into a strain of mice which are free of the virus. This is a case, rather rare in the animal-virus field, of the use of an 'indicator host'. In plant-virus work, however, the use of such indicators is the routine method of testing for the presence of latent virus infections. Several plant viruses can, however, be put into this second category, in which the symptoms disappear after the initial reaction, including the viruses of tomato blackring and others which cause ring-like lesions on the leaves.

To the third category belong those viruses which apparently never cause a disease in their original host, and cannot be induced to do so. In this group are mainly plant viruses, such as the latent infection of dodder, *Cuscuta* sp., and some soil-borne viruses which frequently occur in certain common weeds.

The fourth category may represent a fundamentally different condition; the original hosts are not known ever to have shown symptoms, and they seem free of detectable virus until they are subjected to some appropriate stimulus, when the virus becomes detectable and kills the host. Examples of this type occur commonly in insects and in the lysogenic bacteria. The 'viruses in search of a disease' which have recently been discovered in the various secretions of man and monkeys may also belong to this category.

We give now a few examples of latent virus infections and we shall then discuss some methods of detecting them. We have said that latent infection is extremely common with all types of viruses and this is particularly true of those attacking plants and insects. Of the former, examples occur in hops, strawberries, potatoes, raspberries, dahlias, sugar beet, weeds of various kinds and even in giant trees of the tropical forests. Recent research has uncovered a host of soil-borne viruses which lurk unsuspected in many common weeds and are only brought to light when they infect crops of economic importance such as raspberries, in which they may produce a serious disease.

Latent virus infections also occur in abundance in the larval forms of insects, particularly in caterpillars; but, as we shall see later, they also lurk unsuspected in quite other kinds of insects in which they may sometimes be induced to show themselves.

Most of the latent virus infections in insects are of the polyhedrosis type and both nuclear and cytoplasmic viruses occur in the latent state. There seems little doubt that large populations of some species exist in which the greater proportion of the larvae carry a latent virus infection. This is particularly true of the garden tiger moth, *Arctia caja*, of the silkworm, *Bombyx mori*, and of the spruce budworm, *Choristoneura fumiferana* Clem., in Canada.

Among the higher animals, mice and other rodents carry numerous latent virus infections, and rabbits carry virus III in the testes. Many of London's pigeons and starlings have a latent infection of psittacosis, whilst a high percentage of the domestic fowls are carriers of the leucosis virus complex, to which group belong the sarcoma and leukaemia viruses; domestic fowls are also capable of carrying the virus of fowl-pox.

Detection of latent virus infections

The presence of latent virus infections can be detected by a variety of methods, the choice of which depends, of course, on the type of virus and host organism concerned. We will commence with the detection of latency in plants. The easiest method is to inoculate the sap of the plant suspected of harbouring a virus into a plant of another species, which reacts to the virus in a manner different from that of the plant under test. This, as already mentioned, is known as an 'indicator plant' and there are many such used in plant virology. The tobacco which reacts to more viruses than any other plant was probably the first to be used as an indicator; it was used to demonstrate the existence, in apparently healthy potato plants, of the ubiquitous virus X, which is carried without symptoms by many potato varieties but which nevertheless substantially reduces the yield. It is therefore a matter of economic importance to growers of seed potatoes to identify those potato plants which are harbouring this virus. Other indicator plants include the French bean *Phaseolus vulgaris*, the cowpea *Vigna sinensis*, and certain chenopodiaceous plants like *Chenopodium amaranticolor* and *Gomphrena globosa*. The last two react to a very large number of different viruses, and have proved of great service in their identification.

Another useful means of testing for the presence of a latent virus infection is the serological technique. Many plant viruses are excellent *antigens*, which, put very briefly, means that if they are partially purified and inoculated into a suitable animal, such as the rabbit, they give rise to an *antiserum* which reacts by forming a precipitate only with that particular virus and its related strains. This technique is useful because it affords a rapid and delicate test for the presence of a latent virus infection, and also because it may reveal unsuspected relationships between a latent infection and other viruses causing unmistakable symptoms.

The serological technique has been used by Kassanis, working at the Rothamsted Experimental Station, to throw some light on what has long been one of the most puzzling problems of plant virology. Many years ago, Salaman and Le Pelley working at Cambridge made the interesting discovery that all potato plants of the variety King Edward VIIth carried a latent virus infection. The presence of this virus was discovered, more or less by accident, when a scion of an apparently healthy potato plant of King Edward was grafted to another potato variety which soon developed a severe virus disease to which the name 'paracrinkle' was given. Later it was discovered that every single plant of the King Edward variety from any part of the U.K., including the mutation Red King, was infected with the paracrinkle virus. The paracrinkle virus is never found in any other potato variety or other plant in nature, which is not surprising since the virus has no natural means of spread, not being seed- or soil-transmitted, and having no insect vector. Therefore the virus must have started in the original plant of King Edward, which was a seedling raised by a railwayman in Lancashire. The interesting question was, therefore, how did the virus get into the original seedling? Because there seemed no obvious answer to this, the paracrinkle virus has been quoted as an example of a heterogeneous virus, that is, one that arose spontaneously in the potato seedling. However, the work of Kassanis, together with new knowledge on the mutation of plant viruses, has suggested another more likely explanation of the problem. He found that the paracrinkle virus is serologically related to two other latent virus infections, potato virus S and a carnation latent virus, of which one, the carnation virus, is aphid transmitted. Another recent discovery, relevant here, is that some insect-transmitted plant viruses lose their power of being insect-borne if they are kept out of contact with their insect vectors for long periods.

The suggested explanation for the paracrinkle problem therefore is this: the virus was carried by aphids from carnations to the original King Edward seedling and once there it lost by mutation its insect-transmissibility, helped no doubt by the continued vegetative propagation of the potato tubers.

The story of paracrinkle does not quite finish here; since some viruses fail to invade the growing point of the plant, the apical meristem may be cut off and grown in tissue culture. When large enough the plantlets can be transferred to soil and a virus-free plant obtained. By this method Kassanis has produced a stock of King Edward potato plants which not only look different from the ordinary stocks but give a higher yield.

However, when the virus-free King Edward plants are grown in the field close to commercial stocks of King Edward, the paracrinkle virus spreads from the commercial stocks to the virus-free plants. Apparently the virus can be carried from paracrinkle plants to virus-free plants of the same potato variety by the aphid but not to or from any other plant. This interesting phenomenon could not be encountered previously, since no King Edward plants free of paracrinkle virus existed.

Stimulation of latent virus infection

Sometimes what is called serial passage in an animal results in the development of symptoms, possibly because a latent virus infection has increased in virulence or has been stimulated to more vigorous multiplication. For example, latent virus infections in mice can be brought to light by serial passage of body-fluids from mouse to mouse, progressive inoculation with lung extracts frequently producing typical pneumonia from which a virus can be isolated. Similar serial transmission can stimulate into activity the virus of Theiler's mouse encephalitis.

Not much work has been done on serial passage in insects and plants. It seems quite probable that there would be some sort of virus response in insects, even if it was only that a latent infection was picked up in a tolerant individual and introduced to one which was intolerant.

There is no very precise information on the effect of serial transmission of latent plant-virus infections, but presumably a tolerant host would continue to be a tolerant host after serial passage through the same plant-species. However, complications can easily arise if the virus is passed through a different plant-host; in some cases the virus may actually be changed slightly, possibly by a slight change in its chemical composition. Such a change occurs during the passage of tobacco mosaic virus (TMV) from the tobacco plant to the French bean; after which it seems to have lost much of its infectivity for tobacco. Sometimes the insect relationships of a virus are changed by passage through, or sojourn in, an unusual host; we saw something like this in the discussion of the paracrinkle problem.

Another effect of an unusual host-plant may be to produce an apparent change in virulence which is probably due to selection of a virus strain from a mixture of strains; some strains will multiply more readily in certain plant-hosts than in others and so will tend to obliterate the symptoms caused by the original strain of virus.

Sometimes it is possible to stir latent virus infections in animals into activity by the use of artificial irritants and chemicals. When methyl-cholanthrene is injected into a fowl which has a latent infection of the Rouse I sarcoma virus, it reacts to form a typical Rouse I sarcoma. In the case of rabbits infected with the Shope papilloma virus application of tar causes the papilloma to become malignant; here presumably the action is in changing the response of the cell to the virus. Similarly the application of hydrocarbons to fowls with a latent infection of fowl-pox virus not only activates the virus, but causes a malignant growth to develop.

The work of Yamafugi in Japan has shown that it is possible to stimulate development of a polyhedrosis in silkworms by feeding them on foliage contaminated with nitrates and other chemicals. There are several instances known of a latent virus infection being stimulated into virulence by the introduction of a foreign virus. This phenomenon must not be confused with the increased virulence (synergistic) which may result from the combined action of the two viruses. In a case of true stimulation the second virus plays no part in the subsequent disease but merely acts as a triggering mechanism. This is well shown by some of the insect viruses, the larvae of many insects carrying latent virus infections. Cross-inoculation studies have shown that inoculation with a nuclear polyhedrosis virus will frequently stimulate into activity a latent cytoplasmic virus infection. For example, no naturally occurring virus disease of the caterpillars of that notorious pest of apples, the winter moth, has ever been observed, but inoculation of a number of apparently healthy winter moth larvae with a foreign nuclear polyhedrosis virus from the caterpillars of the Painted Lady butterfly caused a high mortality. Examination of the dead larvae showed that they had not died from the nuclear virus with which they had been inoculated but from a cytoplasmic polyhedrosis. This virus once stimulated into action was easily transmissible to healthy winter moth caterpillars.

This phenomenon opens up some rather interesting possibilities, since it should now be possible to produce virus diseases in unusual types of insects in which virus infections have not previously been observed. For example, until recently there was no record of a polyhedrosis or indeed of any virus disease in insects belonging to the order Neuroptera which includes insects known as lace-wing flies or golden eyes (Hemerobiidae). As is well known, the larvae are predacious and feed mostly on aphids; they will, however, also feed on other insects either alive or recently dead. It has been observed recently that when larvae of Hemerobiids are fed upon the cadavers of *Lymantria dispar* infected with a nuclear polyhedrosis, a varying proportion of them die of a very similar polyhedrosis within 10 days. In spite of the similarity of the two diseases, there are differences which suggest that this is not a case of cross-infection but of stimulation into activity of a virus latent in a proportion of the lace-wing fly larvae.

Examples of the stimulation of latent virus infections by the action of another virus also occur in the higher animals. In man one of the commonest cases of this phenomenon is the so-called 'cold sore' in which the virus of herpes simplex is stimulated into activity by the presence of the common cold virus.

When the papilloma virus from the cottontail rabbit is inoculated to domestic rabbits a papilloma is produced in which no virus can be demonstrated. However, if the papilloma virus is mixed with the virus of sheep dermatitis and then inoculated to the domestic rabbit, papillomas are formed from which active infectious papilloma virus can be recovered.

Environmental conditions, such as temperature and light intensity, have a very great influence on the development of symptoms in plant-virus diseases, and on the rate of virus multiplication, but there does not seem to be any information on the actual stimulation of a latent plant virus by environmental conditions.

On the other hand, adverse environmental conditions sometimes stimulate latent virus infections of animals into obvious activity. For example, parrots when crowded together under dirty insanitary conditions frequently suffer from an outbreak of psittacosis. Crowding is frequently a stress factor which evokes latent virus infections, and this is particularly true of the caterpillars of lepidoptera.

We have seen that the phrase *occult virus* has been recommended for use in place of 'masked virus', about which there has been some confusion. 'Masking', especially in plants, may follow simply from too small a virus concentration, whereas an occult virus is one which is known by circumstantial evidence or by a series of indirect tests to be present but which is not of itself directly demonstrable. One example of such an occult animal virus is that of rabbit papilloma already briefly referred to. This virus causes warts on the skin of cottontail rabbits in the middle western states of the U.S.A., and it is easily transmitted in series indefinitely in the cottontail. When inoculated to the domestic rabbit similar warts are formed but no virus is transmissible from them either to cottontail or domestic rabbits. There are, however, indirect means of proving that the papilloma virus is present in the domestic rabbit tumours; these methods are immunological, and make use of the virus neutralization test or the demonstration of active immunity.

The virus of swine influenza is another example of an occult virus. Here, as we have seen previously, is a reservoir host mechanism which keeps the virus going for some nine months of the year between epidemics. Virus cannot be detected by direct means in the larval lungworms or in the intermediate earthworm host or in the adult lungworm after transmission to the pig. Swine which have become parasitized with lungworm known to be carriers of the occult virus do not come down directly with swine influenza. They must have some provocative stimulus, and the most effective consists of injections of a bacterium *H. influenzae suis*. Curiously enough this provocation of the occult influenza virus only succeeds under experimental conditions between September and April.

A condition in plants which could be more truly referred to as virus masking can be brought about by growing at high temperatures plants infected with certain viruses. Under these conditions the mosaic mottling of the leaves then disappears and the plant appears normal; on returning to a lower temperature, however, the symptoms reappear. It is possible that this may be due to a reduction in virus concentration, brought about by the adverse effect of high temperature on virus multiplication. In the case of *Abutilon* mosaic the bright yellow mottling disappears and is replaced by the normal dark green if the plants are kept in the dark for a time.

It seems, therefore, that in considering latent virus infections there are two main conditions. There is first the occult virus as apparently occurring in the swine influenza and the rabbit papilloma; in this condition the virus itself seems to be in a different physical state and to be in a non-infectious condition. In the other conditions of latent and inapparent infections, these seem to be purely the result of a virus–host equilibrium and the virus itself differs in no way from the same virus in another host which is intolerant of the infection and develops overt symptoms.

10

TUMOUR VIRUSES

The title of this chapter at once stimulates the question: are viruses the cause of cancer? We, can perhaps, put the question in a slightly different form, and ask: what kinds of cancer are known to be caused by viruses? We shall discuss the answer presently.

For some reason there has long been, and still exists, a strong opposition to the theory of a virus etiology of cancer, and some workers persist in referring to 'tumour-agents' rather than viruses as if there was some fundamental difference between a tumour virus and an 'ordinary' virus. In this context a statement by Huebner is apposite:

> Why is the concept that cancer in man may be due to viruses such a difficult one to accept? A virologist who has been in this field for any length of time can hardly think that family relatives of the numerous tumour viruses of animals will not find expression in the human species. To say that such a virus has never been demonstrated is quite correct. However, it is equally correct to say that the critical experiments which have been necessary for the demonstration of animal cancer viruses have not yet been performed in relation to cancer in man.

The fact that no virus has been demonstrated as yet in a human cancer does not necessarily mean that it is not present; it may be there but in insufficient quantity for detection. It is pointed out by Beard that as it has taken fifty years to identify the virus of erythromyeloblastic leukosis in so simple a medium as chicken blood plasma, the fact that viruses have not yet been demonstrated in human cancers is not really surprising.

The more we learn about the versatility of viruses, their mutability and, on the whole, their lack of specificity, the less inclined one is to be dogmatic on the all-important subject of tumour viruses.

During recent years advances in techniques, to which we have referred earlier in this book, have also helped to advance the study of viruses and of their connexion with malignancy. One such technique is the development of the tissue culture of human cells. Another is the application of electron microscopy and the cutting of ultrathin sections to the study of tumours. To these may be added the new methods of electron 'staining' by which the virus particles are thrown into prominence, thus making easier their identification in their sections of cells.

We will now discuss a few of the many tumours in organisms other than man, in which viruses are known to be the cause, or at least one of the causes. The first step in this direction was made by Ellermann and Bang in 1908 when they discovered the virus of chicken leukosis, now considered to be a complex of viruses causing neoplasms in fowls. This discovery, like that of Iwanowski, the discoverer of the first virus, passed almost unnoticed. It was followed in 1911 by Rous's discovery of the sarcoma in chickens, which is now known all over the world as the Rous sarcoma. The idea that a tumour could be caused by a filter-passing agent, in other words a virus, was bitterly opposed in many quarters, and the suggestion was even made that single cells passing through a defective filter-candle were the real cause of the tumour. However, no one now, in view of the number of similar phenomena, doubts the existence of the Rous sarcoma virus, though it may still be referred to occasionally as an 'agent' or a 'factor'. During the following two and a half decades no less than eighteen virus sarcomas of birds were described.

Perhaps the next most important discovery was that of mammary cancer in mice, which was shown by Bittner to be caused by a virus. This was the first time a carcinoma of virus origin in a mammal had been found. Bittner made it clear that the genetic background and the hormonal make-up of the mice also played a part in the formation of the mammary tumour. Thus, there are low cancer-incidence and high cancer-incidence strains; the virus is normally transmitted in the mother's milk, so that foster-nursing could alter the rate of tumour-incidence according to whichever foster-mother was used. That the virus is actually transmitted to the offspring in the mother's milk is well shown by the following experiments. Andervont removed the young mice from the mother by Caesarian section and thus prevented any absorption of milk from the high cancer-rate mother. By doing this he suppressed entirely the mammary cancer which, in the line of mice being used, had an incidence of 97 per cent. Conversely, when the virus was introduced into a cancer-free line the incidence of the disease rose to over 70 per cent. Dmochowski points out that it was characteristic of the attitude prevailing at that time (1936) that pressure was brought to bear upon Bittner to call the mammary tumour virus a 'milk factor' and not an agent, as 'an agent' implied a virus.

In 1933 the papilloma virus affecting the wild cottontail rabbit of North America was discovered by Shope. The disease in the wild rabbit consists of benign tumours or papillomas which only occasionally develop into malignant tumours. However, when inoculated into the domestic rabbit the virus gives rise to papillomas which almost invariably become cancerous; nevertheless the infectious virus has not yet been demonstrated in these tumours.

Tumours have also been observed in cold-blooded animals, the most significant being a kidney tumour in the frog (*Rana pipiens*), described by Lucké in 1934. These tumours frequently spread through the body (=metastasize) and appear in the bladder, intestine, ovary, peritoneum and sometimes in the lungs; they can be transmitted by grafting and have been transplanted into toads and young salamanders.

Another transmissible malignant disease of virus origin is one of the mouse leukaemias; this is especially relevant to our discussion because it has been quite recently shown by de Harven and Friend that the disease is induced by means of a cell-free filtrate and, moreover, the virus itself has been characterized on the electron microscope.

Great interest has been aroused by the work of Gross, Stewart and Eddy, and others on a tumour virus in mice called the 'polyoma virus' which has remarkable properties. It was given this name by Stewart and Eddy because of its ability to form many different kinds of tumours in various strains of mice; the virus can induce as many as twenty-three different types of tumour, affecting, among other parts, the lung, thyroid, pleura, kidneys, adrenals, skin, stomach, thymus, mucous glands and mammary glands. Furthermore, the virus can extend its range to other species and will produce tumours in rats and hamsters. It has been shown by Dmochowski and his co-workers that, in addition to forming tumours of the liver and lungs and subcutaneous tumours in hamsters, the polyoma virus will induce haemorrhagic diseases when inoculated into newborn hamsters.

During recent years much information has been obtained on the various neoplasms of chickens, generally classified as the chicken leukosis complex; these were the first tumours shown to be induced by viruses and they include the Rouse sarcoma. Only one of these various neoplasms is actually contagious: the virus causing visceral lymphomatosis which can be transmitted through the egg, in the saliva and faeces and through the drinking-water. It is also apparently very variable and under certain conditions can cause many different types of tumours.

There are several leukaemia diseases in mice and they seem to be of different origin. They are important not only because of some similarities to leukaemia in man but because, as we have already seen, some at least are transmissible by means of a cell-free filtrate.

In the contribution to the study of tumour viruses made by the electron microscope, the mouse leukaemias play a prominent part. De Harven and Friend have investigated the leukaemia which was induced by a cell-free filtrate, and they have examined particularly the fine structure of the virus itself and the virus–host cell relationship. The virus particles are spherical or near-spherical in shape; their structure corresponds essentially to the concentric spherical shells or membranes, the diameters of which average 87 mμ for the outer and 52 mμ for the inner one. De Harven and Friend have observed what they interpret as distinct stages in the formation of the leukaemia virus. It appears that the first recognizable stage is an increase in the density of the cell membrane, forming a little bulge. During this stage of the differentiation, a small dense crescent appears beneath the bulging membrane, corresponding to the future inner virus shell. The growth of this bulge forms a protruding virus; though still attached to the cell membrane by a pedicel, the cell membrane, the pedicel and the future virus outer membrane are continuous, and the inner membrane of the virus is progressively completed. Narrowing of the pedicel probably liberates the mature particle into the extracellular spaces.

A surprisingly large number of ultrathin sections of tumours, when examined in the electron microscope, appear to contain viruses or virus-like particles. Howatson and Almeida have made such a study of the polyoma virus in the kidneys of hamsters. The kidneys were taken at daily intervals after injection of a variant of polyoma virus into newborn animals. Five to six days after injection, virus particles were seen in great numbers; the most numerous particles measured about 28 mμ in diameter and were similar to the particles seen in large numbers in polyoma-infected mouse cells growing *in vitro*.

De Harven and Friend have also observed in a Swiss mouse lymphoma particles of a type different from those they found in the mouse leukaemia. They point out, however, that the relation of these particles to the lymphoma has not yet been proved.

Fawcett has described virus-like particles in ultrathin sections of the Lucké frog carcinoma and Dmochowski and his co-workers have also demonstrated virus-like particles in mouse leukaemias.

Oberling and his colleagues have studied the internal structure of the virus-like particle associated with the Rous sarcoma and other chicken tumours.

Benign Shope fibroma cells and those of its malignant form in newborn rabbits, have revealed in the electron microscope characteristic changes, and what appear to be various stages in the development of virus-like particles.

Thus we see that the electron microscope reveals the presence of virus-like particles in a great variety of animal tumours and malignant diseases. However, it is not enough just to see such particles in the tumour; the connecting link between the particles and the tumour must be found. In other words Koch's postulates must be fulfilled and the particles seen must be shown capable of producing the disease. This, of course, has been done in several cases: in one of the mouse leukaemias, for example, where the disease is produced by means of a cell-free filtrate and the growth of the virus particles observed in tissue-culture cells; Stewart, also has induced tumours by means of a cell-free extract of polyoma virus. Similarly the gradual formation of virus particles of Shope fibroma and the various developmental stages have been described in cells of rabbit fibroblasts grown *in vitro* and infected with a cell-free filtrate of the Shope fibroma virus.

It must be admitted that so far not much progress has been made towards the demonstration of a virus connexion with human cancer, but as we pointed out at the beginning of this chapter not many critical experiments have been made. Nevertheless, there are some indications which are suggestive; thus the work of Schwarz has demonstrated the presence in the brains of human leukaemic subjects of something which activates leukaemia in mice long before they would otherwise have shown the disease. Bostick also presents evidence for the virus etiology of Hodgkin's disease, and has isolated what appears to be a new virus. It is obvious, however, that this is only a beginning and that a great deal of fundamental work remains to be done.

Two characteristics of viruses are particularly relevant to our discussion: one is the very common phenomenon of latent infection and the other is the mutability of viruses. As we have already shown in the preceding chapter latent virus infections occur in all types of living organisms and man is no exception. Over the last few years an ever-increasing number of 'viruses in search of a disease' have been isolated from the human body, indeed no less than fifteen have recently been described. Presumably some of these viruses must be capable of having effects of some kind, and the production of a tumour might well be one of them. Furthermore, we have already pointed out how variable in their behaviour some viruses are; and how in some cases a virus can be induced to give rise to malignancy. Duran-Reynals has shown that if methycholanthrene is applied to the skin of a fowl infected with the virus of fowl-pox the result is the appearance of skin neoplasia. Again there is the classical work of Rous and his co-workers in which it was demonstrated that the application of coal-tar to rabbits affected with the Shope papilloma virus induced cancers much sooner and more frequently than if the virus was acting alone.

Similarly, it has recently been shown that giving small doses of X-rays to some mice will activate a latent virus to form lymphoid tumours, and carcinogens and sometimes other viruses will activate a latent polyoma virus in mice to form tumours.

As we have already stated opponents of the theory of virus etiology of cancer have expressed doubt as to whether the tumour-inducing agents are viruses and consider they should be in a class by themselves. Is there then any real difference between 'infectious' and 'tumour-inducing' viruses? Eddy points out that except for its capacity to induce tumours the polyoma virus of mice has all the general characteristics of many well-known viruses. It can be freed from other viruses by various methods and, at least, partially purified by adsorption and alcohol-precipitation methods. It can also be characterized by its cytopathic effects in tissue culture and by complement fixation.

Both types of viruses are dependent upon the living cell for their multiplication, and morphological, immunological, and biochemical studies have failed to reveal any essential differences. The isolation- and purification-methods for tumour-viruses do not differ markedly from those used for 'infectious' viruses and, as Dmochowski points out, the purification of the papilloma virus of rabbits was successfully accomplished many years before the isolation of such 'infectious' viruses as that of poliomyelitis and the Coxsackie virus.

In discussing the question as to whether cancer viruses are different in some fundamental respect from ordinary viruses, Stanley asks whether it is possible that a cancer virus is what it is by virtue of a special integrative property involving attachment to and multiplication with the chromosomal apparatus. Alternatively does it operate as a cancer virus by virtue of some type of growth-restraint that holds the growth of virus within certain bounds but stimulates the cell to great growth? Some light may be thrown on these questions by the possibility of growing several tumour viruses in tissue culture.

It is a well-known phenomenon in virology that some viruses when acting together in the host, both plant and animal, may have an interfering effect on each other and one virus may even suppress the multiplication of the other. The possibility of using this antagonistic effect of viruses occurring together has been envisaged as one approach to the cancer problem. It is of interest, therefore, to read of the recent isolation of a tumour-destroying virus; this has been reported by Graham Bennette from the Courtauld Institute of Biochemistry at the Middlesex Hospital Medical School. The virus, which appears to be a new one and has been characterized on the electron microscope, was isolated from certain tumours in mice (mouse ascites). The virus solution when injected into the mouse begins to act within 15–20 hr and in many instances the tumour is effectively destroyed. When the tumour is totally destroyed as the result of early treatment, the animals are indistinguishable from normal mice. One feature of the virus is that it seems completely harmless to the mouse itself, even when inoculated in doses far exceeding those necessary to destroy the tumour.

So far we have discussed only those viruses which induce tumours in the higher animals, and it may be worth while, before concluding the discussion, to inquire into virus tumours in other types of organisms. A few tumours, all apparently non-malignant, have been recorded in insects, in the fruit-fly, *Drosophila*, and in the stick insect, *Dixippus*, for example, but the etiology of these tumours is obscure. However, a non-malignant virus tumour has been described by Bird in the European spruce sawfly, *Gilpinia hercyniae*. The tumours occur in the mid-gut of the larva and consist of a necrotic pigmented mass of epithelial cells surrounded by thin tissue-layers formed from the small proliferating cells.

Although there are several viruses which stimulate growth in plants, there is only one example of a tumour virus which could be considered comparable to tumour viruses in animals. This is the wound tumour virus affecting leguminous plants which has been investigated in the U.S.A. by Black and his colleagues. The name was given because of the marked tendency of the virus to form a tumour at the site of a wound, however slight. Thus, tumours were more frequent on the roots because breakage of the root hairs during the growth of the roots was sufficient to stimulate production of a small tumour. The outgrowths are genuine tumours and consist of completely unorganized tissue; when propagated in tissue culture the tumours continue to grow in size but do not give rise to any type of organized growth.

We have seen earlier in this chapter how the application of methylcholanthrene to the skin of a fowl infected with fowl-pox virus induces skin neoplasia and how the application of tar to rabbits affected with the Shope papilloma virus caused the development of cancers much more rapidly than if the virus was acting alone. It is very interesting then to find a similar type of response in a plant-tumour virus; Black and his co-workers have shown that alpha-naphthalene acetic acid acts synergistically with the virus to stimulate tumour formation. (Where the combined action of two agents in a single host is greater than either acting alone, it is called synergism.) The augmentation of tumour growth can be observed histologically within 72 hr after the application of the chemical to the apical nodes of infected plants.

11

CONTROL OF VIRUS DISEASES

Animal viruses. Plant viruses

Animal viruses

The methods of attempting to control the virus diseases of man and the higher animals can be grouped broadly into three categories, *avoiding* or *preventing infection*, *immunization procedures* and *chemotherapy*.

Avoiding infection

Under this heading come a number of miscellaneous measures designed to avoid or prevent the spread of virus infection. One of these is the attempt to destroy the arthropod vectors of a particular virus; this has been made more efficacious by the invention of new, more persistent insecticides, such as DDT. This has been most effective in anti-mosquito campaigns and is almost ideal for the control of *Aëdes aegypti*, the vector of the yellow fever virus, when applied as a residual spray to the walls of houses and to the outside and inside of water containers. DDT has also been used with success against the fly *Phlebotomus papatasi*, the vector of phlebotomus fever.

Modern air transport has brought with it additional hazards in the possibility of infective virus vectors travelling in aeroplanes from one continent to another. Such a possibility is the introduction of mosquitoes infected with yellow fever virus into India; a technique for the thorough disinfestation of trans-continental aeroplanes is therefore necessary.

In the case of some tick-transmitted viruses, it is important to avoid the tick-infested areas or to wear suitable clothing which prevents the ticks gaining access to the body. With some vectors or suspected vectors it is difficult to take any preventive measures; it has long been suspected that starlings may carry on their feet the virus of foot-and-mouth disease from the Continent into the United

Kingdom. No measures to prevent this seem possible at the moment, short of concerted efforts to exterminate a pestilential bird.

The isolation of patients suffering from infectious virus diseases is an obvious course of action, as is the quarantining of animals entering a country. The United Kingdom has been free of rabies for many years, thanks largely to the rigid quarantine laws affecting the importation of dogs, and when there has been an isolated case of this disease it has been due to the criminal action of some person smuggling a dog into the country. In many states of the U.S.A., however, rabies is endemic, and this is combated by the mass-immunization of dogs, stray-dog control and strict quarantining. Presumably the muzzling of all dogs in an affected area would also be enforced.

Unfortunately in some districts in the U.S.A. and in other countries, there are additional sources of rabies infection. Epidemic rabies is sometimes common in wild animals such as wolves, foxes, coyotes, jackals and skunks, and this has to be combated by shooting and trapping. In hot countries such as South America, Trinidad and parts of the U.S.A. an unusual winged vector of rabies has appeared; this is not an insect but a blood-sucking or vampire bat which feeds indiscriminately on the blood of man and cattle. Since the bat attacks man only when he is asleep this danger can be countered by the screening of windows. It is a different matter, however, to protect cattle, and in Mexico more than two million head of cattle have been immunized against the disease.

In the case of some animals which are capable of carrying a dangerous virus disease, the embargo is used in place of quarantine. This means that no such animals are permitted to enter the country; the embargo has been used more than once against the parrot, because of the danger of psittacosis or parrot fever. We have seen elsewhere how pets can be potential reservoirs of viruses dangerous to man, and parrots should never be allowed into close contact with him.

The slaughter policy is used with two very infectious virus diseases of animals; to be effective this must be completely ruthless and not only the infected animals but all their possible contacts must be slaughtered and burned. This policy is used to try to stop the spread of foot-and-mouth disease in cattle, etc., and fowl-pest in poultry and other birds. At the time of writing (1960), more than 18,000 head of cattle have been destroyed by slaughter and burning. In 1959 more than 1,500,000 birds were slaughtered because of fowl-pest at a cost of over £1,250,000 in compensation.

Influenza is essentially a 'crowd disease' and it is obviously safer to avoid, if possible, places such as cinemas, or football crowds, during an influenza epidemic. The attempted sterilization of the air in buildings by means of aerosols or ultraviolet light does not appear to be effective.

Immunization procedures

The classical case of immunization is, of course, vaccination against smallpox introduced by Jenner in 1796; in this case a mild strain of the virus is used, cow-pox virus or vaccinia from which the inoculation procedure gets its name. Vaccination is now a general term and is not confined to immunization against smallpox.

Vaccination can be of three kinds, using: (1) inactivated or 'killed' virus; (2) active or 'live' virus; (3) passive immunization in which large quantities of antibodies from a recovered person or animal are used.

In all these methods, the objective is to call forth in the vaccinated person or animal the antibodies or immune bodies which confer protection against a particular disease without producing any serious ill-effects.

Great steps forward have been made in recent years in the development of vaccines against serious virus diseases, and this has largely been made possible by the advances made in the tissue culture of viruses, in which the name of John Enders in the U.S.A. is outstanding. There are two vaccines which have stood the test of time and have become routine inoculation procedures. The first of these is, as we have already mentioned, vaccination against smallpox, which was at one time compulsory and should be made so again. The other is against yellow fever, and both these vaccines contain an active but attenuated virus. There are two strains of attenuated yellow fever virus used for immunization; one is the French neurotropic strain which has been adapted to the mouse; the other is a safer, less virulent form known as 17D which has been produced by prolonged cultivation of the virus in tissue culture.

Thus we see that virus mutation, occurring either naturally, or induced by experiment, can be of great service to man. There is, however, the other side of the medal, and this is well illustrated by the behaviour of the influenza virus; in 1957 there appeared the Asian strain of influenza A virus which immediately flared up into a pandemic since it was different antigenically from all other known strains and found everywhere a population susceptible to infection. The natural mutation of viruses is therefore one of the great difficulties against producing vaccines and what is wanted is a 'master vaccine' which would be effective against the various mutant strains.

In addition to the above two vaccines containing active attenuated viruses, vaccines containing inactivated or 'dead' virus have been successfully used against the three important diseases, epidemic typhus, influenza and poliomyelitis.

In the following short account of the situation regarding vaccination against several common virus diseases I am indebted for much of the information to *Viral and Rickettsial Infections of Man* (3rd ed.), edited by T. M. Rivers and F. L. Horsfall.

Against rabies there is no specific treatment once the disease has developed, but two types of inactivated virus vaccines are used. One is rabbit-brain tissue infected with the Pasteur rabbit-brain fixed virus which has been inactivated by incubation at 37° C; the other is a similar virus but inactivated by ultraviolet light. As a preventive of the spread of the virus mass-immunization of dogs is carried out with the Flury LEP modified rabies 'live'-virus vaccine. This gives an immunity for at least three years.

One of the great advances in the immunization against dangerous virus diseases was made by Salk and his co-workers, who produced a vaccine against poliomyelitis. Virus grown in tissue culture of monkey kidney-cells and inactivated by formalin was used for this vaccine. Inactivation by formaldehyde permits reliably reproducible results and retains the property of giving rise to antibodies. Attempts to inactivate the virus by means of ultraviolet light were unsuccessful, since doses sufficient to inactivate the virus also destroyed the power to produce antibodies. No doubt in the future continuously propagated cultures of cells will make the use of monkey kidney cells unnecessary.

Against influenza much good work has been done in the development of vaccines, the great difficulty, as already pointed out, being the occurrence of so many different strains of the virus. In the U.S.A. in 1943 subcutaneous vaccination with concentrated inactivated virus which had been cultured in chick allantoic fluid was found to have an effectiveness of about 75 per cent against epidemic influenza A. The effect of the vaccine began to be noticed in 7–10 days after inoculation, when the antibodies were accumulating. In 1945 the same vaccine which contained both type A and type B influenza viruses was shown to be effective in preventing an epidemic of influenza B. However, in 1947, this same vaccine was ineffective in the face of another epidemic of what was called the A-prime strains. This illustrates the need of developing a 'master' vaccine.

In 1957 an intensive effort was made in the U.S.A. to provide a vaccine against the Asian strain of influenza which was seen to be approaching. Material for the start of vaccine studies was obtained in two months, and the preliminary results suggested a 40–75 per cent protection. In the period of August to December over 50 million doses of vaccine were distributed, a remarkable achievement.

Studies on the immunization against influenza using an attenuated or 'live' virus have been mostly carried out by Soviet scientists.

Measles, which was first clearly identified as a separate entity by Sydenham in the seventeenth century, is one of the most contagious of all diseases. It is mainly a disease of childhood, but it can affect adults; although not usually fatal, it is nevertheless responsible for as many deaths as whooping-cough or poliomyelitis. Until recently immunization against measles was of the passive type only, consisting of doses of immune bodies from recovered persons. However, it has now been announced from the U.S.A. that a measles vaccine has been successfully used in a small group of children. The vaccine, which contains active attenuated virus, produced an effective response in all the children. There is still some way to go, of course, before this vaccine can come into general use; there is always the danger in the use of a 'live' vaccine that the attenuated virus might revert to a virulent form.

One of the most annoying, if not the most serious, virus diseases with which man has to contend is that perennial plague, the common cold. It has also proved a most difficult and refractory problem to isolate the causative virus. At the common cold research centre in Salisbury, C. H. Andrewes and his co-workers have studied this problem with the aid of many human volunteers. Now, after thirteen years of research, it has been announced that a virus has been obtained from three individuals with 'wild' colds, and the material continues to produce colds in volunteers after as many as eight tissue-culture passages.

Success was finally achieved by growing the virus in embryonic human kidney tissue at a temperature of 33° C, which is some degrees lower than is usual and in a slightly more acid medium. In two cases the cold virus had a characteristic effect on the cells (cytopathic); this is important because it shows the presence of the virus without having to test for it in volunteers.

Trachoma is the greatest single cause of blindness in the world, and it is rampant throughout the Far East, the Indian subcontinent, Asia and Africa. Although recognized as a virus for many years, it was first isolated in 1957 by Chinese workers, and this was confirmed shortly afterwards by workers at the Lister Institute in London. The virus is of large size, resembling those of psittacosis and lymphogranuloma venereum, and it can now be cultivated on the yolk-sac of the developing chick embryo. Now that it can be cultivated in tissue culture in quantity, it should be possible in the near future to attack the disease from a new angle.

Passive immunization consists of large doses of immune serum from persons who have recovered from a particular virus disease and so have antibodies circulating in the blood. This treatment must be given before exposure to the virus (prophylaxis) or at least before the appearance of symptoms, since the antibodies cannot affect the virus once it has started to multiply inside a cell. This type of immunization has been used with some success during measles epidemics, but it is likely to be supplanted before long by the attenuated virus vaccine to which we have already referred.

Chemotherapy

It has been a disappointment, if not perhaps a surprise, that the antibiotics and sulphonamides have proved useless for the treatment of diseases caused by the small viruses. They are of some use, however, against the larger agents such as those causing psittacosis and trachoma.

The ideal chemotherapeutic agent would be one which was capable of selectively inactivating the virus precursor materials. In other words what is wanted is a substance which interferes with the production of the nucleic acid of the virus. It is here, of course, that a very high degree of specificity would be required, otherwise the normal production of nucleic acid in the actively reproducing cells of the body might also be inhibited. As Sir Alexander Todd has pointed out in his Jephcott Lecture: 'It is this need for a very high degree of specificity, coupled with the need for complete and irreversible inactivation of certain polynucleotides that makes the problem of virus and tumour diseases so intractable.'

Before concluding this short account of the control of animal-virus diseases, there are one or two other lines of approach which might perhaps be included under the heading of chemotherapy. One of these is the discovery of a substance to which the name 'interferon' was given by Isaacs and Lindemann in 1957 because it interfered with the development of the virus. Interferon was discovered during the investigation of a well-known phenomenon in which one virus, when it has occupied a cell, interferes with the growth of other viruses in that cell. It retards virus multiplication in susceptible tissues but does not act directly on the virus. In discussing whether interferon can be used in much the same way that vaccines can be used to exploit the body's antibody-forming defence mechanisms, Isaacs and Hitchcock point out the following advantages of interferon as a potential antiviral agent. First, it is not toxic in doses which prevent viral growth; secondly, it is not antigenic, so that it should be possible to use it repeatedly without any resistance being built up to its action. Thirdly, it has a wide spectrum of action, being effective against most of the viruses tested and, fourthly, it is active in the animal body, preventing, for example, the development of lesions due to vaccinia virus in the skin of the rabbit.

A discovery which may have a great significance in the fight against tumours is the recent announcement by Graham Bennette of the isolation of a tumour-destroying virus. The agent, which in the electron microscope has all the appearances of a virus, has a destructive action on mouse ascites (tumours). The solution containing the virus begins to act within 10–20 hr of being injected into the animal, and in many instances the tumour is effectively destroyed. When, as a result of early treatment, the tumour is totally destroyed, the treated mouse is indistinguishable from healthy mice. Furthermore, doses far in excess of that necessary to destroy tumours seem to have no ill-effect upon healthy mice.

Finally, there is the recent announcement of a new chemical which is apparently efficacious against influenza, mumps, measles and one or two other virus diseases. No further details of this chemical seem available at present.

Plant viruses

Whilst it is obvious that the main approach to the control of animal- and plant-virus diseases must differ fundamentally, there are nevertheless one or two aspects of the subject which have factors in common. It must be remembered that plants do not recover from virus diseases, and the necessary farming-practice of growing huge numbers of plants in close proximity naturally favours the spread of infection. Our first category in the control of animal-virus diseases was avoiding infection and this type of control is also applicable to plant-virus diseases. The direct attack on the arthropod and other vectors of plant viruses has not so far been very successful, largely because it is necessary to kill a very high percentage of the vectors, since a small number is quite sufficient to spread the disease. However, the development of the systemic insecticides, that is, those which are taken up by the plant itself, has helped to lessen the spread of some plant viruses. This is particularly true in the case of the 'persistent' viruses; these are viruses which can be picked up by the vector in one feeding period, which thereafter remains infective for long periods without access to a fresh source of virus. The important point here is that the vector has to feed for a comparatively long period before it can become infective and is thus killed by the insecticide before it can move off and infect a healthy plant. This is in contrast to the 'non-persistent' viruses which can be picked up by the vector in a few moments and distributed to fresh plants before death of the vector ensues.

Avoiding the vector and the elimination of sources of infection are methods which are also applicable to plant viruses. The growing of 'seed' potatoes in certain areas of Scotland is an example on a big scale of avoiding the vector, since the aphid, *Myzus persicae* Sulz., which is chiefly responsible for the spread of potato viruses, cannot thrive in the moist climate of the 'seed'-growing areas.

In Germany, efforts are made to avoid growing potatoes in the vicinity of peach trees; this is due to the fact that the aphid *M. persicae* overwinters on peach trees in the form of a winter egg. In spring, therefore, these trees act as a prolific source of aphids.

Sometimes it is possible to avoid a bad infestation of the aphid vector by early sowing. This is recommended in the case of the sugar beet and the disease of virus yellows; by early sowing the crop will be well advanced and more in a position to withstand a subsequent aphid attack.

Elimination of the sources of virus infection is not easy with plant viruses. Obvious examples are the removal of 'volunteer' potatoes or sugar beet which have been left over from the previous year's crop and which experience has shown are frequently infected with virus. Another example is the removal of the seed or mother beet plants from the vicinity of the sugar beet root crop; this is done because the seed plants are biennial and frequently infected in the autumn with virus yellows. Moreover, the aphid vector overwinters on these plants and so is ready in the spring to fly to the new season's beet crop bringing the virus with it.

Immunization, which plays such an important role in the control of animal-virus diseases, is hardly applicable to plants since they do not give rise to antibodies. The nearest thing to an acquired immunity to virus infection in plants is a non-sterile type of resistance which holds good only for related viruses. In other words if a plant is infected with one strain of a virus it cannot usually be infected with another strain of the same virus. This only becomes of practical application when it is possible to infect a crop deliberately with a mild strain which will then protect the plant against a more virulent strain of the same virus. So far this method is mainly of academic interest.

Of recent years more attention has been given to eliminating a virus from a plant by means of heat without destroying the host. A pioneer in this heat therapy was the American virologist L. O. Kunkel, who first showed that peach trees infected with the viruses causing peach yellows, little peach, rosette, etc., could be cured of these diseases by subjecting them to a temperature of 35° C. The trees were kept at this temperature for a fortnight or longer, and the time necessary was longer for the larger trees. Now this treatment has been applied to a number of different viruses and their host plants and plants affected with about thirty viruses can now be cured by heat therapy; these include potatoes, strawberries, cherry trees and sugar cane. The control of the ratoon stunt disease of sugar cane by hot-water treatment is now practised on a large scale. In Queensland in 1953, over 2000 tons of cane setts were exposed for 2 hr in hot water at 50° C; this was carried out in wire baskets holding a ton at a time, immersed in special tanks.

The chemotherapy of plant-virus diseases is only just beginning, and it is not easy to prophesy its future. The principle underlying the chemical treatment is that multiplication of a virus can be delayed by compounds which interfere with nucleic acid metabolism. Since the purine and pyrimidine bases are the important part of nucleic acids, and confer their specificity upon them, attention has been paid to possible virus-inhibitory agents among synthetic analogues of the natural bases. Matthews found that 8-azaguanine, when sprayed on to plants, had a quite marked effect on the spread of virus within the plant. It was found effective against one or two viruses, the reason being that the base was incorporated into the nucleic acid and thus rendered a proportion of the virus particles incapable of initiating infection.

If it is desired to resuscitate or rehabilitate, a valuable plant or plant variety, this can sometimes be done by special methods of propagation. For example, by taking advantage of the rate of movement, or lack of multiplication of a virus, in certain tissues, it may be possible to build up a virus-free clone of a particular variety. In the case of dahlias infected with the virus of tomato spotted-wilt, it is possible, at a time of rapid growth, to take cuttings from the tips of shoots as they arise from the tubers. The virus fails to keep pace and there are often a few inches of tissue not yet reached by the virus. The potato King Edward has long been known to carry a latent virus infection, called paracrinkle, and no plant of that variety has been found free of it. In consequence, no one knew what a virus-free potato plant of this variety looked like. Now, since some viruses fail to invade the growing point, the apical meristem can be cut off and grown in tissue culture; when large enough the plantlets can be transferred to soil and a virus-free plant obtained. By this means Kassanis has produced a stock of King Edward potato free of the paracrinkle virus: these plants not only look different but give higher yields than the infected plants.

Finally, there is the method of breeding virus-resistant or immune varieties of plants. Some good results have already been obtained by this method; there are the mosaic-resistant varieties of sugar cane, sugar beet resistant to the curly-top disease and cotton strains which are resistant to the leaf-curl diseases.

In producing a potato plant resistant to some viruses, a condition known as 'field-immunity' is aimed at. This means that the plant is so susceptible to the virus that it is killed outright; in this way the virus is not only destroyed but the possibility of its spread to healthy susceptible plants is also eliminated.

12

VIRUSES AS AGENTS OF BIOLOGICAL CONTROL

Effects of Virus Diseases. Obtaining and Applying the Viruses. Attempts at Control. Future Possibilities

The idea of trying to control insect and other pests of agricultural crops by 'setting a thief to catch a thief' is not a new one. Biological control, as it is usually called, has been used widely, not always with success. The earliest efforts in this direction employed mainly predacious or parasitic insects, which were produced in large numbers under laboratory conditions and then introduced into the area where the pest in question was active.

In the early days of biological control there were three large-scale attempts to eradicate certain insect pests by the introduction of predacious or parasitic insects. In 1888 the vedalia beetle, *Rodolia cardinalis* (Muls.), was introduced from Australia into California to control the cottony-cushion scale, *Icerya purchasi* Mask., on citrus. In 1874 a ladybird beetle, *Coccinella undecimpunctata* L., was sent from England to New Zealand to try to keep down aphids. Then in 1883 the hymenopterous parasite, *Apanteles glomeratus* L., was also sent from England to the U.S.A. in an attempt to control the caterpillars of the white butterfly. Of these three experimental introductions the importation of the vedalia beetle was the most successful, and the cottony-cushion scale is no longer a serious menace to the citrus industry.

Biological control, however, is not confined to the encouragement of predatory insects, and more attention is now being paid to the possibility of starting epidemics of disease among agricultural pests. Both bacterial and fungal disease agents have been tried, and some success has been obtained with a bacterial disease of the Japanese beetle, *Popillia japonica* Newn., by producing the spores of *Bacillus popilliae* Dutky, the causal agent of the milky disease, in large quantities and disseminating them in the soil against the larvae in the eastern United States.

The idea of using viruses in biological control is also not new, the classical example being the introduction of the myxoma virus into Australia against the rabbit pest and, more recently, into France and Great Britain. What is new is the use of viruses to control insect pests, and this is not surprising, since it is only during the last decade that insect viruses have been seriously studied. For some reason these agents have been neglected by virologists, which is a pity, because they are of great interest and quite suitable for fundamental studies on the nature of viruses.

Effects of virus diseases

As mentioned earlier, the distribution of virus infections in the insect kingdom appears to be very uneven, but this may be merely a reflexion of our lack of knowledge. The greatest number of viruses, mostly causing polyhedral diseases, undoubtedly occurs among the larvae of the Lepidoptera (butterflies and moths), as anyone who has tried to rear caterpillars will agree. There are several virus diseases of hymenopterous larvae (sawflies), and two have been recorded from the larvae of Diptera (two-winged flies). No authenticated case of a virus disease has been recorded in the Orthoptera (grasshoppers, locusts and cockroaches), Hemiptera (aphids and plant bugs), or Coleoptera (beetles), though no doubt viruses exist in these orders and will be discovered sooner or later.

The manifestations of virus diseases in insects seem to depend very largely on the tissues affected. Thus, in the nuclear polyhedroses and in the granuloses the skin is attacked and rendered excessively fragile; it ruptures at a touch and liberates the liquefied body contents, including the polyhedra or granules, over the food-plant. With the cytoplasmic polyhedroses, and where the virus is free in the tissues, the skin is not attacked. When these larvae die the body becomes flaccid and eventually dries up, unless, as frequently happens, secondary invasion by bacteria hastens the destruction of the body. In caterpillars that have died of a cytoplasmic polyhedrosis, white patches are often visible through the skin; these are masses of polyhedral crystals situated in the mid- or hind-gut.

Obtaining and applying the viruses

It is of course necessary, first of all, that a virus specific for the insect must be found. In other words, except under special circumstances, which are dealt with later, it is no use applying any insect virus and hoping it will attack the particular pest. How the virus is to be found is merely a question of looking for it. The procedure is simple enough and consists in the systematic examination of all dead or apparently diseased individuals of the insect to be controlled. The polyhedral diseases are easily identified, and the routine test consists in making a blood smear of the larva on a slide, fixing it with heat, and staining with Giemsa's solution. Under the oil-immersion lens of the optical microscope the polyhedral crystals are readily detected. But where the disease is caused by a virus free in the tissues and there are no intracellular inclusions, identification is more difficult and an electron microscope must be used.

Having obtained the necessary virus the next step is to produce it in sufficient quantities for use in control. Since it is a fundamental property of viruses that they cannot multiply outside a living cell, it follows that the virus must be propagated in the insect itself. Large quantities of the larvae must therefore be raised and infected artificially with the virus. In the case of the polyhedroses and granuloses the bodies are collected and stored in water contained in large conical flasks; these are left standing at room-temperature for some weeks to allow the bodies to disintegrate and the polyhedra or granules to sediment to the bottom of the flask. The debris is then decanted and the virus material collected by centrifugation. A stock of the infective material can be kept indefinitely.

Polyhedral and granular viruses are the most suitable for use in biological control, for they are easily identified by means of the optical microscope and, more important, the occlusion of the virus itself inside a protein crystal not only protects the virus from environmental conditions but allows it to be sprayed in a viable and convenient form. In nature the polyhedral crystals assist the spread of the disease in a remarkable manner: they are extremely resistant to cold and other factors and may retain their infectivity for years. Some are carried by the wind, while others remain on the food-plant to infect later broods of larvae.

Spread of insect viruses in the field is mainly by ingestion of virus-contaminated food-plants and by inheritance from infected parents. There is little doubt that many of the insect viruses are passed through the egg to subsequent generations, and this, of course, is of great significance in the matter of control. It is thus possible to get a virus seeded into a population of insects and affect future generations in a way no insecticide can do. Contamination of the food-plant is also important and is closely bound up with the virus and the type of disease induced by it.

The nuclear type of polyhedral disease and the granuloses attack the skin and reduce the body contents to a semi-liquid material containing millions of polyhedra or granules. The skin ruptures and the liquid contents run down the stem of the plant or are splashed around by rain. Moreover, these liquefied cadavers can be attractive to the uninfected larvae, which feed upon them with disastrous results. Cytoplasmic polyhedral viruses do not have this liquefying effect upon the body, nor do they attack the skin, so the polyhedra are not so easily spread. This is partly compensated for, however, by the fact that large quantities of polyhedra are voided with the faeces and contaminate the food-plant.

Application of the virus in the control of the insect pest follows the methods used for spraying insecticides. Preliminary trials may be required to get a rough idea of the number of polyhedra per ml. necessary for good control; usually the number is about a quarter to half a million, and the polyhedra can be conveniently estimated by the use of a device for counting blood cells.

The normal procedure is for the polyhedra to be suspended in water and applied with a low-volume sprayer at the rate of 10–15 gallons to the acre. Sometimes it is necessary to add a 'sticker' of some kind to make the polyhedra adhere to the leaves, but as a rule they stick quite well and are not easily washed off by the rain.

Feeding-habits are also important in determining if an insect will be amenable to this kind of control. The greatest success is likely to be attained with insects that feed openly and gregariously on the leaves of the plant; insects that live in the soil or feed hidden inside a stem or bud are much more difficult to reach with a virus suspension.

Attempts at control

Some of the pioneer work in this form of control was done in the U.S.A. One notable attempt was made in California against a serious pest of lucerne, the caterpillar of the clouded yellow butterfly (*Colias philodice eurytheme* Boisd.). The caterpillars feed openly on lucerne and are easily reached with a virus suspension. A polyhedral virus was used and an initial supply was built up by infecting caterpillars in the insectary; this was then used to spray a field containing a high population of caterpillars. By collecting the infected larvae in the field with a sweeping net on the day before they are expected to die of the disease, large quantities of the virus may be obtained: in one of the field tests where approximately half a pint of the suspension was applied, 7 gallons of virus material were recovered in 4 hours' time on the sixth and seventh days after treatment. Spraying was done both by aeroplane and by ground equipment. The conclusion was drawn that more extensive trials are necessary before it can be said that this method of controlling the lucerne caterpillar is an economic success. Timing of the application is extremely critical, because once damage begins to show in the field it is too late to use the virus as a practical means of control. In fields with very high populations it may be necessary to apply the virus before the caterpillars hatch from the eggs.

Another interesting experiment was carried out in Canada against the European pine sawfly, *Neodiprion sertifer* (Geoffr.). The larvae of this sawfly resemble caterpillars in some ways and feed gregariously on the foliage of the Scotch pine. The insect was first reported in Canada at Windsor, Ontario, in 1939 and by 1949 had spread throughout most of the south-western part of that province. No virus disease had been observed in the sawfly populations in Canada, but a polyhedral disease of the same insect was known in Sweden. The virus was imported into Canada and propagated until enough was available for testing on a large scale; it was then sprayed over the forest by means of an aeroplane. Mortality from the disease was found to depend on the amount and concentration of virus used, the method of dissemination, and the stage of larval development at the time of spraying. To produce the greatest mortality the virus should be applied at, or soon after, the time of hatching of the eggs, but serious defoliation could be prevented by treatment before the larvae reached the fourth instar.

Similar experiments on the large-scale use of a polyhedral virus to control defoliating larvae are being carried out in South Africa by Ossowski. The insect concerned, *Kotochalia junodi* (Heyl.), is known as a 'bagworm' because it lives inside a case or bag made of plant material or debris. It is indigenous to South Africa, where it lives on species of *Acacia*, from which it has invaded plantations of black wattle (*A. mollissima* Willd.)—a crop of considerable economic importance.

Aqueous suspensions of 10,000 polyhedra/mm^3 caused a very high mortality which was not increased when greater concentrations were used. Such suspensions remain effective for a considerable time, so that they can be applied some months before the bagworms hatch. This may ensure a high mortality of the newly hatched larvae and thus save time in large-scale applications. There seems to be no danger that the polyhedral bodies will be washed off the trees by rain, for Ossowski relates that during one experiment a total of 550 mm of rain fell without apparently impairing the efficacy and distribution of the polyhedra.

Experiments on a smaller scale are being carried out in Great Britain. As a preliminary to this work, viruses were found which would attack about twenty species of injurious insects in Britain. Many of these viruses are not suitable for practical use and some of the insects are not of sufficient economic importance to warrant the experimental effort, but work is being carried out on the control of the larvae of the following pests: the two white butterflies *Pieris brassicae* and *P. rapae*, the pine sawfly *Neodiprion sertifer*, and two species of the clothes moth, *Tineola bisselliella* (Hummel) and *Tinaea pellionella* (L.).

A granulosis virus was used against the caterpillars of the white butterflies. This virus attacks the skin of the larvae and reduces the body contents to a liquid mass, so that the highly infectious 'granules' are readily spread. The caterpillars are extremely vulnerable to attack by means of a virus suspension, for they feed openly and in large numbers on low-growing brassica crops which are easily reached with a spray. The granulosis virus used is highly infectious and virulent, so much so that five fully grown caterpillars in an advanced stage of the disease are sufficient to make one gallon of spray. The virus suspension is made up in water, and a 'sticker' is added to make it adhere to the waxy brassica leaves. It must be applied as early in the attack as possible, but the effect is fairly rapid and infected larvae stop feeding some time before the disease becomes manifest. Unfortunately for the purposes of experiment, serious attacks of these caterpillars occur only at intervals, and the opportunity to try out the virus on a large scale has not yet been forthcoming.

Immediate practical results have been achieved against the larvae of the pine sawfly on a plantation in Norfolk, where a local but severe outbreak of the pest occurred. Half the plantation was sprayed with a suspension of polyhedra and the other half left unsprayed as a control. The virus has not been recorded in England, though it does occur in Scotland, and its introduction seems to have been very successful, for a mortality of about 80 per cent was obtained. Moreover, when the plantation was visited again in the following year, it was found that the virus had moved across to the unsprayed part of the plantation, and again a heavy mortality had resulted. How the virus got into the control plantation in such a high concentration is not quite clear; it seems unlikely that the wind could have carried over sufficient polyhedra. A probable explanation is that infection was transmitted through the eggs of sawflies arising from larvae that had picked up the virus in a late stage of their development. Such larvae may have failed to develop the disease yet have retained some virus throughout the metamorphosis. It is to be hoped that the virus will now become established in the local populations of sawfly larvae and that periodic outbreaks of the disease will be sufficient to keep the pest under control.

The methods used for the control of the clothes moth are much the same as the foregoing. Suspensions of polyhedra are sprayed on to garments and in or around the cupboards and drawers. The main difficulty in the way of the successful use of this virus on a large scale is the small size of the larva and the low concentration of virus obtainable in consequence.

Future possibilities

It will be clear from this chapter that the use of viruses in the control of insect pests is only just beginning and that much more investigation into the possibilities is needed. Indeed, each attempt to apply the method is a separate piece of research in which all the variable environmental factors governing this type of biological control must be taken into consideration. It may be permissible, however, to speculate a little on some possible future developments. We have seen that in several large Orders of insects no viruses have ever been discovered, and the same applies to individual pests belonging to Orders in which viruses have been recorded. What are the possibilities of inducing virus diseases in insects that have never been known to be thus infected? This is a fascinating problem and one which has hardly been investigated. It can be approached from two aspects: attempts can be made, first, to take advantage of the phenomenon of latent virus infections which are widespread in the insect kingdom and, second, to infect the pest with a virus from a different insect host.

It is an interesting fact that large populations of insects carry latent infections, usually of the cytoplasmic polyhedral type. These can sometimes be stimulated into activity by stress factors such as overcrowding and excess humidity. Frequently, too, a latent virus infection can be 'triggered off' by inoculation with a foreign virus. A case in point is the caterpillar of the winter moth, *Operophtera brumata*, which has never been found with a natural virus infection. When apparently healthy caterpillars of this moth were fed with a *nuclear* polyhedral virus from the caterpillar of the Painted Lady butterfly, *Vanessa cardui*, a cytoplasmic polyhedral disease developed, and this, once stimulated into action, could be transmitted to other winter moth larvae and a supply of the virus built up. This is an instance of stimulating a latent virus in an insect belonging to an

Order in which many viruses occur, but we can envisage the application of the method to other Orders of insects with no record of virus infection.

The second approach to this problem is to try to infect a given insect with a foreign virus, and not just stimulate a latent infection. The axiom has been laid down that all insect viruses are species-specific; but axioms are liable to hold up progress in scientific research, and this one is no exception. Although it is true that a virus may be highly specific when attempts are made to transmit it by feeding to other insect species, this may not hold if the virus is injected into the insect by means of a micro-needle. Viruses have been transmitted by this method from one insect to another in which no virus disease had previously been described.

In chapter 8 we have seen that viruses have been discovered in other members of the Arthropoda, that is, the red spider-mites. This means that a new chapter in the control of a formidable group of pests has now been opened, and experiments in the use of a virus in the biological control of the citrus red mite are being carried out in California.

What is badly needed in the United Kingdom and Europe generally is a virus to destroy the fruit-tree red spider. This mite has now become a first-class pest of apple trees; there are two reasons for this, one is the destruction of its natural enemies by the indiscriminate use of insecticides and the other is the development of races of mites which are resistant to the acaricides now in use. A virus attacking the European fruit-tree red spider has now been discovered in America, and it is hoped that it may eventually be introduced into Europe.

As research proceeds viruses are discovered in new hosts and one of these, mentioned in chapter 8, is a suspected virus disease of a nematode worm, another serious pest of agricultural crops. Whether it will be possible to develop this method of control for plant-feeding nematode worms remains to be seen.

LITERATURE

Chapter 1

Mayer, Adolf *et al.* (1942). *Phytopathological Classics.* No. 7. (Amer. Phytopath. Soc.)

Smith, Kenneth M. (1957). *Beyond the Microscope.* (Penguin Books.)

Williams, Greer (1960). *Virus Hunters.* (London: Hutchinson and Co.)

Zinsser, Hans (1937). *Rats, Lice and History.* (London: George Routledge.)

Chapter 2

Burnet, F. M. and Stanley, W. N. (1959). *The Viruses* (series): *Plant and Bacterial Viruses* (vol. 2); *Animal Viruses* (vol. 3). (New York and London: Academic Press.)

Rivers, T. M. and Horsfall, T. L. (1959). *Viral and Rickettsial Infections of Man* (3rd ed. London: Pitman Medical Publishing Co.)

Smith, Kenneth M. (1960). *Plant Viruses* (3rd ed. Methuen's Biol. Monogr.)

Sonneborn, T. M. (1959). Kappa and related particles in *Paramecium. Adv. Virus Res.* **6**, 231. (New York and London: Academic Press.)

Chapter 3

Bawden, F. C. (1950). *Plant Viruses and Virus Diseases.* (3rd ed. Waltham, Mass. U.S.A.: Chronica Botanica Co.)

Burnet, F. M. and Stanley, W. M. (1959). *The Viruses* (series): *General Virology* (vol. 1). (New York and London: Academic Press.)

Smith, Kenneth M. (1951). *Recent Advances in the Study of Plant Viruses.* (2nd ed. London: J. and A. Churchill.)

Chapter 4

Williams, R. C. (1954). Electron microscopy of viruses. *Adv. Virus Res.* **2**, 183. (New York and London: Academic Press.)

Wyckoff, R. W. G. (1958). *The World of the Electron Microscope.* (Newhaven, U.S.A.: Yale University Press.)

Chapter 5

Enders, J. F. (1959). Tissue culture techniques employed in the propagation of viruses and rickettsiae. *Viral and Rickettsial Infections of Man*, ed. by T. M. Rivers and F. L. Horsfall. (3rd. ed. London: Pitman Medical Publishing Co.)

The Nature of Virus Multiplication (1953). Second Symp. Soc. Gen. Microbiol. Oxford 1952. (Cambridge University Press.)

The Nature of Viruses (1957). *Ciba Foundation Symposium*. (London: J. and A. Churchill.)

Westwood, J. C. N. (1959). Tissue culture in relation to viruses. (Current Virus Research.) *Brit. Med. Bull.* **15**, 181.

Willmer, E. N. (1958). *Tissue Culture*. (3rd ed. Methuen's Biol. Monogr.)

Chapter 6

Gordon Smith, C. E. (1959). Arthropod-borne viruses. *Brit. Med. Bull.* **15**, 235.

Shope, R. E. (1959). The natural history of hog cholera. *Perspectives in Virology*, p. 145. (London: Chapman and Hall.)

Chapter 7

Broadbent, L. and Martini, C. (1959). The spread of plant viruses. *Adv. Virus Res.* **6**, 94.

Harrison, B. D. (1961). Soil-transmission of plant viruses. *Adv. Virus Res.* **7**, 131.

Reeves, William C. (1958). Arthropods as vectors and reservoirs of animal pathogenic viruses. *Handbuch der Virusforschung*, p. 177, ed by. C. Hallauer and K. F. Meyer. (Vienna: Springer.)

Smith, Kenneth M. (1958). Arthropods as vectors and reservoirs of phytopathogenic viruses. *Handbuch der Virusforschung*, p. 143, ed. by C. Hallauer and F. K. Mayer. (Vienna: Springer.)

Smith, Kenneth M. (1960). Relationships of plant viruses with their arthropod and other vectors. *Plant Viruses*, p. 40. (3rd ed. Methuen's Biol. Monogr.)

Chapter 8

Bergold, G. H. (1958). Viruses of insects. *Handbuch der Virusforschung*, p. 60, ed. by C. Hallauer and K. F. Meyer. (Vienna: Springer.)

Smith, Kenneth M. (1959). The insect viruses. *The Viruses* (Series), III, 369, ed. by F. M. Burnet and W. M. Stanley. (New York and London: Academic Press.)

Chapter 9

Recent Progress in Microbiology. Latent and Masked Virus Infec ion Symposium IV. VIIth International Congress for Microbiol. (Stockholm, 1958.)

Smith, Kenneth M. (1952). Latency in viruses and the production of new virus diseases. *Biol. Rev.* **27**, 347.

Symposium on Latency and Masking in Viral and Rickettsial Infections (1958), ed. by D. L. Walker, R. P. Hanson and A. S. Evans. (Minneapolis, U.S.A.: Burgess Publishing Co.)

Chapter 10

Andervont, H. B. (1959). Problems concerning the tumour viruses. *The Viruses* (Series), III, 307, ed. by F. M. Burnet and W. M. Stanley. (New York and London: Academic Press.)

Beard, J. W., Sharp, D. G. and Eckert, E. A. (1955). Tumor viruses. *Adv. Virus Res*, **3**, 149.

Current Medical Research (1959). Cancer-producing viruses and their immunology. *Rep. Med. Res. Counc.* 1957–8, p. 24.
Dmochowski, L. (1958). The part played by viruses in the origin of tumors. *Cancer*, **1**, 214.
Dmochowski, L. (1959). Viruses and tumors. *Bact. Rev.* **23**, 18.
Dmochowski, L. (1960). Viruses and tumors in the light of electron microscope studies: a review. *Cancer Res.* **20**, 977.
Huxley, Julian (1957). Cancer biology: viral and epigenetic. *Biol. Rev.* **32**, 1–37.
Rous, Peyton (1959). Surmise and fact on the nature of cancer. *Nature, Lond.* **183**, 1357.
Rous, Peyton (1960). The possible role of viruses in cancer. *Cancer Res.* **20**, 672.
Stanley, W. M. (1960). Virus-induced neoplasia-outlook for the future. *Cancer Res.* **20**, 798.
Stewart, Sarah E. and Eddy, Bernice E. (1960). The polyoma virus. *Adv. Virus Res.* **7**.
Viral Neoplasia. *Perspectives in Virology*, ed. by Morris Pollard. (New York: John Wiley and Sons.)

Chapter 11

Current Virus Research (1959). *Brit. Med. Bull.* **15**.
Rivers, T. M. and Horsfall, T. L. (1959). *Viral and Rickettsial Infections of Man.* (London: Pitman Medical Publishing Co.)
Smith, Kenneth M. (1960). *Plant Viruses*, p. 186. (3rd ed. Methuen's Biol. Monogr.)

Chapter 12

Smith, Kenneth M. (1960). Some factors in the use of pathogens in biological control with special reference to viruses. *Report Seventh Commonwealth Entomological Conference*, p. 111. (London: 56 Queen's Gate.)

INDEX